eBook Bonus.

With the purchase of this book, you also
receive access to its electronic version.
Use the following personal code to download
the electronic pages.

70188-r62lg-
ur400-en3l2

Take advantage of the eBook Bonus on
your computer, smartphone, or tablet. Go to
www.hanserpublications.com/ebookbonus
and register with your personal access code.

Patrick Tarvin

Leadership & Management of Machining

Patrick Tarvin

Leadership & Management of Machining

How to integrate technology, robust processes, and people to win!

Hanser Publishers, Munich   Hanser Publications, Cincinnati

The Author:

Patrick Tarvin is a successful manufacturing leader who has demonstrated a proven ability to turn-around machining organizations in multiple industries. He has held positions in large and medium organizations including plant manager, engineering manager, supervisor, and machinist.

Distributed in the Americas by:
Hanser Publications
6915 Valley Avenue, Cincinnati, Ohio 45244-3029, USA
Fax: (513) 527-8801
Phone: (513) 527-8977
www.hanserpublications.com

Distributed in all other countries by:
Carl Hanser Verlag
Postfach 86 04 20, 81631 München, Germany
Fax: +49 (89) 98 48 09
www.hanser-fachbuch.de

Cataloging-in-Publication Data is on file with the Library of Congress

Print-ISBN 978-1-56990-639-2
E-Book-ISBN 978-1-56990-640-8

© 2016 Carl Hanser Verlag München
www.hanser-fachbuch.de
Editor: Dipl.-Ing. Volker Herzberg
Production Management: Cornelia Rothenaicher
Typesetting: Kösel Media GmbH, Krugzell
Coverdesign: Stephan Rönigk
Printed and bound by Hubert & Co, Göttingen
Printed in Germany

Dedicated to:

Donald and Patricia Tarvin
and
Nancy, Jessica, Joe, Robin, Zach

CONTENTS

ABOUT THE AUTHOR

With more than three decades of expertise in the business of machining, Patrick Tarvin constructs a blueprint for machining organizations of all types to achieve success and growth. He completed the machinist training program from one of the all-time great machine tool builders – Cincinnati Milacron. He financed his engineering education as a machinist and went on to earn an MBA and complete numerous Lean and Six Sigma programs. He has been a customer and supplier of machined components at both large and small organizations. He has led numerous acquisitions and turnarounds of machining facilities, and he has held every management and engineering position from machinist to plant manager. He is a machining industry expert who has proven experience integrating technology with lean processes and traditional machining fundamentals to create strong, agile, and winning organizations, a veteran of the aerospace, medical, automotive, and capital equipment industries. His exposure to nearly all types of machine processes, all levels of volume, push/pull, and five-ERP installations, provides the platform for Mr. Tarvin to bridge theoretical business practices with the real world problems faced by machining organizations in a hyper-competitive global economy. His passion is to help Western machining companies succeed and anchor a revitalized manufacturing sector based on innovation and technology.

INTRODUCTION

While some manufacturing processes simply disappear or glacially creep forward due to changing technology, machining has displayed more than a century of evolutionary and revolutionary progress. Just as manual machine tools gave way to NC machine tools, which led to CNC machine tools, the machining industry is now entering its next generation. Only this time, the technological advancements are as much outside the machine tool as within.

The intent of all technological advancements in the machining universe is to increase productivity and precision with less involvement of the endangered species known as the machinist. Connectivity, data, and digitization are providing opportunities for significant improvements in modeling, programming, simulation, machine/people productivity, and real-time decision making. Advances in chip technology and processing speeds yield more sophisticated controls. Software enhancements create advanced CAM systems, which interpret more complex models and produce higher precision programs.

This book demonstrates how to integrate these technological advancements with Lean /Six Sigma, CNC programming, quality, and machining best practices to create an agile machining organization, capable of winning against competition anywhere in the world. Technology by itself is not the answer. It is an enabler.

This book is not about cutting speeds or vibration analysis. It is about the business of machining and how to develop the infrastructure and people to provide not only profits and growth, but long-term sustainability.

Whether machining is just part of your business or "all" of your business, this book is for those at all levels of the company who must develop the machining strategy or execute the machining strategy.

While most publications are written for only executives, I have deliberately designed this publication for the entire organization. The role of everyone on the organizational chart or in the shop is discussed and reviewed. The premise of the book is that the demands of global competition on the modern machining organization are so great that success requires a foundation of leadership, people, equipment, quality, IT, robust processes, and all employees executing to yield exceptional quality, delivery, and profits. Since I examine the soft skills, hard skills, software, hardware, and all the business processes that winning machining organizations must integrate, I do not go a mile deep into any single topic. Other authors have created books for Lean, Six Sigma, supply chain, information technology, etc. I provide the reader with an overview of these topics and over how the varying types of machining companies should employ them to create the architecture and blueprint for high performance.

There are experts in every field that I discuss who possess extreme knowledge in their specific area of focus. There are companies, mostly large organizations with extensive resources, which already successfully apply some or many of my recommendations. Most experts ignore or minimize other functional areas of the machining organization outside of their niche. What I present is a holistic approach to real world machining problems, gained from decades of servicing the most demanding customers in multiple industries. There is no replacement for remaining in a leadership position for five-plus years at a machining facility and living through the successes and failures that result from the decisions you made. I have had this opportunity on four occasions in four different industries: capital equipment, tool and die, medical, and aerospace. Leaders hire, develop, and mentor technical teams. They lead teams that acquire and install complex equipment, processes, and software, and they answer to unhappy customers, suppliers, and employees. It is easy for consultants and academia to preach zero inventory, but when the supply chain falters, it is the plant manager who is responsible for enacting recovery plans, it is the plant manager who is responsible for fixing the problem, it is the plant manager who forces overtime, it is the plant manager on daily calls with the customer, and the plant manager is responsible for poor P&L due to expediting costs and schedule disruptions. These experiences develop the wisdom to decide when to carry inventory, whom to hire, how to develop teams, how to motivate people, when to create lean cells, how to improve CPk, how to manage a supply chain, and how to focus on customers.

Human learning, technology, and productivity have surged throughout history, as new forms of communication have been introduced: writing, printing, photography, radio, TV, and the internet. Will connectivity through the Internet of Things (IoT) and Industry 4.0 be the next catalyst? Through these mediums and with our personal experiences we all stand on the shoulders of those who came before us. I am no exception. For all those listed in my bibliography and all my peers throughout my career, who will recognize many of the topics and solutions I discuss—thank you!

If you are in the machining business, you are running two races at the same time. You are running the 100-meter sprint at this very moment to deliver to your customer this week, and you are running a marathon to develop the people, software, and processes your customers will require in the coming years. Let us get started.

1 PYRAMID

I believe that machining is the most challenging type of manufacturing process from a technical and managerial perspective. Contributing to this is a lengthy era of rapid change, constant pressure to innovate in order to reduce cost, and erratic global markets. From a technical perspective, the high number of input variables that must be controlled to produce complex geometries, precision tolerances, and flawless surface finishes exceeds the demands from other manufacturing processes. A die or mold will produce thousands and possibly millions of components with no appreciable wear. Assembly processes, by nature, have no wear and little variation other than the variation driven by the lower-level components. Conversely, during machining the cutting tools begin to wear on the first part, and there may be dozens of cutting tools on any given operation. Developing economies consistently begin with assembly, fabrication, and continuous process industries (chemical, food) because of these facts. Only later do they begin repetitive machining of high-volume components with simple geometry and open tolerances. Low-volume or batch-volume machining of medium to high complexity is still the domain of advanced economies in Western Europe, North America, and Japan. Often overlooked are the infrastructure requirements to produce critical aerospace, medical, and energy-related products. In low-cost regions, the integrity of the raw materials, heat treat, and chemical coatings is a grave concern.

Manufacturing has evolved in Western developed economies. We started with craft production, moved to the replaceable components of Eli Whitney, through the assembly lines of Henry Ford, and finally onto the Toyota Production System, and now, mass customization. Over the last two decades, machining, or more accurately, Computer Numerical Controlled (CNC) machining and its family of support technologies, have further evolved to the level that we need a new approach to management and technology implementation.

The ability to collect and share real-time data, store and retrieve mass amounts of process documentation, and advances in CAD/CAM accompanied by game-changing metrology and machine tool developments have created a data-connected factory far surpassing productivity and quality potential of past years. Gone are the days of problems building undetected and, once detected, waiting for a team to be assigned until the problem could generate a large enough savings to justify resolving. The new model is people, equipment, and data to create robust processes on day one and the breadth of talent and detection to sustain and continuously improve processes 24/7. I call this management system for modern machining organizations the machining pyramid (see Figure 1.1).

The Machining Pyramid that I am going to discuss is a holistic approach to organizational success for a manufacturing company, with specific emphasis on machining. It is an integration of many business and technical practices combined with my 30 years of success in diverse machining facilities. Many organizations fail to achieve expected results despite management changes, consultants, lean implementation, capital investments and other initiatives. Understanding the Machining Pyramid will display the management system and infrastructure that must be in position to optimize and connect upstream and downstream machining activities.

We learned about pyramids as children. We drew pyramids, we saw them on TV and in cartoons, and most of us have made human pyramids at one time. We certainly still see them today formed by cheerleaders at basketball games. More importantly, we know that the ancient Egyptian pyramids remain well preserved and standing strong after several thousand years.

What you have never seen at the basketball game is the top cheerleader placed in position prior to all the base layers. It is highly probable that when drawing a pyramid you will draw the base layer first. It only seems to be in business that we seek to build the pyramid from the top. In manufacturing, especially machining, we need to build from the bottom and not the top. How many times has someone in your organization stated, "we need to improve our on-time shipments", or "we need to reduce scrap", or the ultimate construction-from-the-top-of-the-pyramid pronouncement, "we need to increase profit margins". The simple reason that these wishes (scrap, on-time deliveries, profits) are upside-down construction is that they are not actions that any given employee or group of employees can perform. They are, in reality, percentages.

Each one is a metric with a numerator and a denominator that represents thousands of outcomes and is influenced by dozens of variables. They are, in fact, by-products of performing other tasks. They are not themselves tasks which can be performed. If you seek improved cost of quality, delivery, or profit, you must perform all the tasks better that aggregate into the numerator or denominator.

For example, profit margin for a given month is an average of the profit for each product shipped during the month. This can be hundreds or thousands of unique components, assemblies, fabrications, etc. The profit on any one item is influenced by the sale price (or for some organizations an internal transfer price) and the total

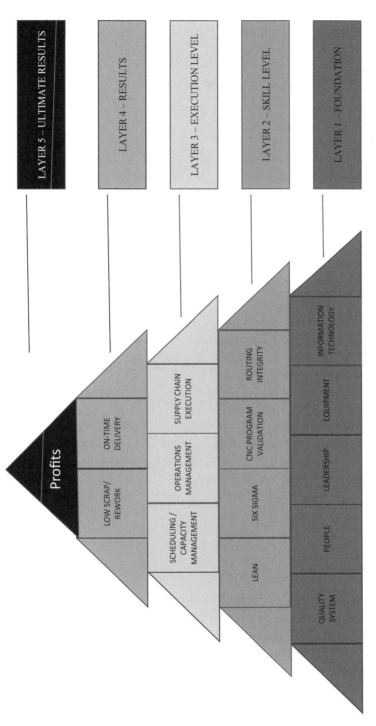

LAYER 5 – ULTIMATE RESULTS

LAYER 4 – RESULTS

LAYER 3 – EXECUTION LEVEL

LAYER 2 – SKILL LEVEL

LAYER 1 – FOUNDATION

Profits

ON-TIME DELIVERY

LOW SCRAP/ REWORK

SUPPLY CHAIN EXECUTION

OPERATIONS MANAGEMENT

SCHEDULING / CAPACITY MANAGEMENT

ROUTING INTEGRITY

CNC PROGRAM VALIDATION

SIX SIGMA

LEAN

INFORMATION TECHNOLOGY

EQUIPMENT

LEADERSHIP

PEOPLE

QUALITY SYSTEM

Figure 1.1

actual cost to produce the item. The sales price can frequently vary and is particularly subject to swings at contract shops and design/build organizations. Total actual cost will vary due to the cost of the material, expediting outside processes, any scrap or rework incurred, and the amount of direct labor at each operation. If your company shipped 100 unique items during the month, and each item required an average of six manufacturing steps (also called operations), then you performed six hundred operations during the month. Each operation was an opportunity to scrap the items or to significantly overrun the expected labor. Each operation also required some level of setup, first piece inspection, tooling changes, and possibly a handoff between shifts. You may have been successful on 575 operations, but the other 25 operations generated enough scrap/rework or additional labor to reduce the profit margin.

Let us look a little deeper at the "tasks" that have to go right to be successful on "each" of the six hundred machining operations performed each month. The machine tool needs to be available and capable of holding the tolerance on the print; the cutting tools, fixtures, and inspection devices need to be located, accurate, and set up correctly; the machinist needs to be properly trained and must not commit errors or spend excess time; the CNC program needs to be accurate and provide enough documentation for the machinist to understand how to set up holders, tooling and fixtures without creating scrap or using excess time; the parts have to be correct from each preceding operation, and in general the overall process created by your engineering or processing group needs to be lean and statistically capable of producing parts within tolerance. All this is challenging when you are machining the same or similar product each month—sometimes referred to as high volume/low mix. For organizations that are producing new components, that have varying quantities and little visibility to customer demand—sometimes called low volume/high mix—managing these tasks is extremely challenging.

As a result, to improve the profit margin you not only need to improve the underlying tasks discussed above, but the tasks must be executed on all shifts, for all operations, on new or old parts, and with experienced or new machinists.

On-time shipments and scrap/rework have completely other sets of tasks which must be executed for each operation, each shift, each day, and each month. These "tasks" are the execution level of building blocks on our Machining Pyramid.

Lean Manufacturing and Six Sigma are an important part of our Machining Pyramid. Your product mix, volume, and overall business model will determine the degree to which you can employ these tools to manage your operations. I believe there exists a separate foundation to your business that, if strong, will allow you to develop and optimize lean tools and Six Sigma tools.

So, our Machining Pyramid begins with a bottom layer that I call the foundation layer (see Figure 1.1). This consists of five strong blocks that, when properly established, will support all organizational activities in good times and in bad times. We are building our pyramid to last for decades. We are building our pyramid to survive earthquakes, storms, and time. Our layers are not just blocks sitting on top of each other. We are building a Machining Pyramid with three-dimensional interlocking blocks. The foundation layer consists of people, equipment, information technology, quality system, and leadership.

There are many ups and downs to owning and operating any business. Manufacturing organizations encounter additional challenges. In addition to the typical economic cycle of boom and bust, the ordinary manufacturing enterprise will periodically lose a key customer, key personnel, or perhaps a key supplier. The manufacturer also periodically must reinvent their products or services, face recalls or serious quality escapes, and other man-made or natural catastrophes. Through it all, a properly instituted foundation layer will provide the strength, wisdom, depth, and breadth essential to negotiate troubles with minimal disruption and stress.

Resting above and interlocked with the foundation layer is the skill layer. These are the tools that will be employed to create processes superior to your competitors, innovative processes that yield better quality and productivity, processes that are easier to manage, simpler, and more reliable. This layer consists of CNC program validation, routing integrity, lean manufacturing, and Six Sigma. In the machining industry, it is not enough to have robust processes that are innovative. All methods and steps must be documented, controlled, and repeatable, so that months or years later the same results can be achieved without additional setup, debug, or scrap. According to the SIC data, there are more than 35,000 organizations performing machining in the U.S. alone. These companies employ diverse business models to satisfy their customer requirements and compete within their niche. I explain how to tailor lean manufacturing and Six Sigma to the various types of machining organizations: original equipment manufacturers, contract machine shops, machining job shops, and high-volume machining.

Now that our foundation layer has provided the human and material elements and the skill layer has created our robust processes, we are ready to manufacture our parts every shift, month, and year. Hence, the next layer is our execution layer. It consists of scheduling and capacity management, operations management, and the supply chain. The execution layer is where we convert the raw material into finished products.

The fourth and fifth layers of our Machining Pyramid are the result layers. If we have been successful on our lower levels, our results will yield low scrap/rework, on-time deliveries, and profits. I will review how to create appropriate metrics to drive improvement further, interpret trends and the relationship of the three result layer blocks.

There will be a chapter devoted to each building block. We will discuss why each of these building blocks is essential, examine the nuances of each, and determine how to measure and optimize for your organization.

A successful company establishes the base layer of their Machining Pyramid as the foundation for success. By executing the functions on the middle layers of the Machining Pyramid your team is positively influencing the individual data points that aggregate into success on the upper layers—low scrap, better on-time delivery, and higher profits.

There are no shortcuts to success in the competitive machining world. You may have some of the building blocks in position, but without your entire Machining Pyramid in place you will limit your growth and profits. The Machining Pyramid is an architectural structure that incorporates hierarchical requirements. A building will not stand with missing columns or missing floors. Your machining operation is the same. Good machine tools without the corresponding metrology equipment or adequate CNC programming will not perform to expectations. People and technology will be thwarted by a lack of scheduling and capacity management capabilities. You cannot be strong in some areas and weak in others—it just does not work in machining.

If our chosen profession was the restaurant industry, we could purchase a franchise and would be provided with a menu, building architecture, dining layout, kitchen equipment, advertising, daily raw-material deliveries as well as training programs, and we would have every business function planned and standardized. There are no franchises in manufacturing, but for machining organizations the Machining Pyramid provides a customizable framework that delivers as close to a franchise as feasible.

LAYER 1
FOUNDATION

2 QUALITY SYSTEM

You will read many times in this book that quality is more than quality. Quality is also on-time delivery, and quality is also productivity. It is a given that if your company does not produce parts to print, you will not be in business very long. Of course, quality can also be aesthetics, service, and adherence to specifications. Why does quality equate to on-time delivery? It is very likely that when you scrap components they will also ship late to your customer, as you normally cannot manufacture replacement parts in time to meet the delivery date. Likewise, when you must rework components they will frequently become late shipments. A quality escape will also be categorized as a late shipment by your customer. More damaging is the reality that scrap parts must be re-manufactured, which will be a series of unplanned production runs on multiple pieces of equipment, consuming your capacity across varying workcenters. These unplanned production runs will delay the completion of other production orders, leading to more late deliveries. If you are unfortunate to have scrap that must be re-manufactured on a bottlenecked workcenter, then you may have created a domino effect of late deliveries.

Why does quality equate to productivity? The most obvious answer is that re-manufacturing or reworking consumes labor and does not generate revenue. Once non-conforming material is uncovered, the entire universe surrounding the questionable material enters slow motion. The organization moves from warp drive to impulse power. Inspection must measure components upstream and downstream to determine the source of the non-conformance and the distance of the escape. Machinists must review their process and measure every component to insure that corrective action has been implemented and that the corrective action was effective. Additionally, most inspection sampling plans require 100% inspection after failures have been located. All these activities bring productivity to a crawl.

It should not be a surprise that quality is a foundation-level building block on our Machining Pyramid. In this context, quality refers to your quality management system, your quality culture, and any quality-related process or program-affecting internal or external quality. I will discuss the quality culture more thoroughly in the Operations Management chapter, and I will discuss the amount and effects of

ork in a separate chapter. Specifically, in this chapter I will discuss each
ow in detail:

1. Borderline Quality
2. Total Quality Management
3. ISO Compliance
4. Quality at the source
5. Geometric Dimensioning & Tolerancing (GD&T)
6. Risk Analysis – Pre-Production Project Review
7. Root Cause and Corrective Action
8. Final Inspection
9. Metrics

BORDERLINE QUALITY

It may be counterintuitive, but there are occasions when scrap is more productive than making parts to print. How can this be? When a part is scrapped, an NCR (non-conformance report) will be created and a root cause implemented to correct the process, so the dimension is machined close to the mean. Future parts will be machined in a productive manner. Problem found and problem solved.

More damaging is the hidden cost and lost utilization of "borderline" quality. When you have non-conforming parts (scrap, rework), you statistically also have parts that are just within the tolerance limits. However, you will also experience many occurrences when parts are just within tolerance limits and do not have any non-conformance on the production order—problem not found and problem not solved.

Six Sigma techniques and statistically capable processes (CPk, Cr) will help avoid these two types of borderline quality (BQ). So why is BQ damaging to productivity and expensive? First, when a machinist is fighting to keep parts within the tolerance limits, he will not allow the machine to run unattended during lunch or breaks. The machinist will not allow the machine to run while he is performing inspections, and he will be forced to drastically over-inspect. While he is performing inspections, he is not performing other activities necessary to keep one or more machines making chips. Secondly, when a machinist is fighting BQ, he will be forced to frequently enter tool offsets and/or frequently change his tooling. Again, this will prohibit the machinist from performing other activities necessary to keep one or more machines making chips. This is also consuming time to change the tools and will be increasing the tooling costs. Third, when the machinist is fighting BQ, he is very likely traveling to or from final inspection and obtaining input from a supervisor or other technical personnel. Again, this will prevent the machinist from performing other activities necessary to keep one or more machines making chips.

A considerable focus of our Machining Pyramid is designed to avoid scrap, rework, and borderline quality. These three villains will be discussed throughout the book. The negative impact of borderline quality is so severe and underestimated that I have broken my own rule and created a new acronym—BQ.

TOTAL QUALITY MANAGEMENT

The highest level of TQM is when the principles become a habit, the culture and the DNA of the company. When this occurs, there is no need for an official TQM program. Every employee at every level of the firm needs to live and breathe quality and integrity at all times. Leadership needs to initiate this expectation during the interview process in order to make it clear to prospective employees that the foundation of the business is constructed on this principle. Instead of asking for suggestions, evaluating suggestions, and determining rewards, the employees and management should just be implementing continuous improvement on a constant basis. If employees believe they will be financially rewarded, they will spend more time writing down the suggestion than implementing the suggestion. Engineers will spend more time evaluating than thinking and implementing. I have been involved in these programs and they degenerate into endless suggestions, trying to earn a small or large payday. How do the evaluators say "no", without making someone angry?

It is really not hard to get everyone working on creating robust processes and insuring that inputs are controlled to produce components to specifications. Employees at every level of the organization have the same goal—do not compete on cost! If you compete on cost, then management needs to restrain compensation and benefits. Nobody wants that scenario. You want to compete on precision, complexity, creativity, quality, reliability, speed, project management, integrity, delivery, and overall customer satisfaction. Reasonable leadership should be able to communicate this point until everyone is drinking from the same Kool-Aid. Besides, error-proof processes are less stressful for the machinists and all the people responsible for producing a world-class product. Employees know they are competing against low-cost countries and it is not a tough sell for everyone to understand that customers are buying confidence and risk avoidance. The customers want their product on the delivery date and they want it to print with no surprises.

Access to real-time data, enhanced metrics, ERP, and the rest of the Machining Pyramid provide transparency and reduce tribal knowledge. Continuous improvement opportunities are revealed. Operations management will ultimately reward and reinforce the behavior and culture through teamwork, empowerment, and career enhancement. TQM philosophy is embedded throughout the Machining Pyramid.

I have experienced union workforces that are hesitant to volunteer improvements. That is an uphill battle. When a non-union workforce is not engaged in continuous improvement, a different management approach is required, and not a suggestion box.

ISO COMPLIANCE

Most organizations are committed to achieving and maintaining one of the many ISO certifications. The most popular are ISO 9001, AS 9100, ISO/TS 16949, and ISO 13485. Of course, the difference is whether you are involved in commercial, aerospace, automotive, or the medical industry. The titles for these certifications and their requirements are updated periodically. It is not uncommon for an organization to hold more than one certification. There are many more similarities between these various standards than differences. These standards establish "what" must be accomplished to meet the standard, but not "how". Whether your organization is required by your customers to be certified to a specific standard, or your organization chose to be certified, you have considerable flexibility to create pragmatic systems that simply represent good business practices.

I am not an expert or advocate for any specific ISO standard. I have implemented and managed facilities with all of these standards. If you do not feel that your quality management system adds value to the organization, then you only have yourself to fault. If you feel your QMS only creates paperwork, you only have yourself to fault.

Your QMS should be focused on prevention. When you identify a potential problem or a non-conformance, a proper preventive/corrective action (level 1 or level 2) will eliminate similar mistakes in the future. It is imperative to apply this lesson to all parts in the family or all parts that utilize the same process (sometimes called "read-across") in order to generate cost savings and continuous improvement. Developmental actions should accompany corrective and preventative actions to shift the organization away from firefighting to pro-active control of your own destiny. A non-conformance needs to be viewed as a learning opportunity. Think of it as a shot across the bow that has prevented a much more expensive or embarrassing mistake. If each opportunity is seized, you will push these savings to the bottom line, reduce BQ cost, and improve on-time delivery.

Your ISO processes and systems should be part of everyone's daily routine and culture. These systems will provide compliance to the standard, but they should also comply with efficiency and integrity. Many of you may be raising an eyebrow, as this seems to be an oxymoron. How do you create or adjust your systems and procedures so that they mirror actual behavior and are efficient? My first recommendation is to restrict quality professionals from writing the procedures. It is easy to write procedures that you do not have to follow or manage. Quality professionals invariably will slant to creating a complex and exhaustive procedure that an auditor will admire. The process owners should have input or preferably write the procedure. I have found that operations management personnel are the most effective at creating the procedures. They understand the time required to perform the tasks, who in the organization has the knowledge and time available, and the relative importance of the task. Operations management is also responsible for enforcement, shop utilization and headcount. Creating an ISO procedure that cannot be realistically followed 24/7 diminishes the entire program, raises integrity concerns, and instigates tension between good people, their managers, and between their managers and the quality organization. Quality professionals frequently write procedures as if they had a blank check. Their perspective can be: "every employee only exists to insure compliance". An effective approach is for the process owner and operations management to create and document an efficient process, while the quality professional identifies gaps with compliance. As a further word of caution, I recommend that operations management actually read the necessary sections of the ISO standard in those cases in which quality wants to use it as a hammer to impose unrealistic procedures. This is an area where effective leadership will play an important role.

Each procedure can become a mini-Kaizen event that should minimize waste. A lean process that employees embrace and amend over time promotes integrity and reinforces the winning culture. Most importantly, these ISO procedures should be part of your electronic solutions kit.

An important aspect of ISO compliance is the audit process. There are generally three types of audits: internal audits, customer audits, and your official registrar audits. Performing internal audits, documenting, and responding to findings are a mandatory ISO requirement. Many employees consider the audit experience to be similar to a visit to the dentist. Unfortunately, dental visits are required to insure long-term health. Audits perform a similar role. The benefit of the audit is the identification of individuals or systems that accidentally or purposefully do not perform as intended. This is important information that allows management to implement corrective action prior to a major quality escape or prior to years of compiling the same mistake. It is better to address the cavity immediately. Audits help to reinforce discipline and promote a culture of living the QMS. Every manager benefits from reminding employees that lapses and shortcuts will be discovered during an audit. Employees do not want to be the reason that an audit is failed or the reason behind an audit finding.

I will take a minute at this time and remind everyone that the reason for procedures of any type and for installing the discipline to adhere to procedures is not for the average employee performing his normal job functions. The reasons for procedures, documents, and discipline are events at 2:30 a.m., new employees, weekends, and employees covering for someone on vacation.

In summary, when you focus on prevention, document your best practices, and leverage audits for instilling a culture of integrity, you will have jointly maintained your ISO certification and optimized daily business practices.

QUALITY AT THE SOURCE

The old adage that you cannot inspect quality into a part refers to measurement of the part at final inspection after it is finished. The goal is to produce quality at the source, which is the machining operation. Quality at the source includes in-process inspection, on-machine inspection, first-piece inspection, CNC program validation, setup and tooling documentation, and all other methodologies that eliminate process variation. You can measure quality into a part while it is being machined, because it is the metrology that creates the correlation between the inputs and the outputs. In fact, you will not obtain a quality part, if you are not proficient at in-process inspection. In-process inspection refers to all of the measurements performed during the various stages of manufacturing—from raw material through the last production sequence. You must understand and control your inputs to have quality at the source. You cannot control the inputs, if you cannot measure and verify your outputs. This requires precision metrology and robust processes at the source. I want to be clear and precise—it is only through exceptional and frequent metrology at the source that you can hardwire the inputs to create a robust, statistically capable, Six Sigma-type process. For those lot sizes and operations that do not reach a quantity threshold for achieving statistical capability, it is the exceptional metrology that permits an organization to economically provide complex and tight-tolerance components.

Different machining organizations are required and are capable of holding different precision levels. Machining tolerances that are easy or "open" still require professional metrology. The market is demanding tighter tolerances and many companies need measurement resolution of a micron or .0001" (see Figure 2.1).

Machining is fundamentally different from manufacturing processes such as assembly, blanking, and injection molding. Each machining setup is "offset" to some degree, depending on tool length, tool radius, tool deflection, fixture location, or size and location of existing part features. These offsets are based upon the in-process inspection measurements or on-machine measurements. The goals are more than just avoidance of scrap and rework. Attainment of "quality" is more than simply adhering to specifications. To surpass your competition and establish a reputation for quality that drives customer decision making, you need three criteria. The first is to produce components with features centered within the tolerance band and with little variation. This is known as maintaining a high CPk or Cr for each feature. These are statistical measures of your capability, with the higher the number the better. The second is superior surface finishes with elimination of machine lines and chatter. And the third is edge breaks and radii that are machine-gener-

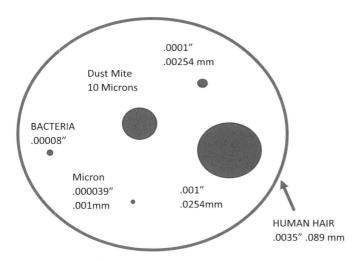

.0001"
.00254 mm

Dust Mite
10 Microns

BACTERIA
.00008"

Micron
.000039"
.001mm

.001"
.0254mm

HUMAN HAIR
.0035" .089 mm

Metrology Accuracy for Machining

Figure 2.1

ated and properly blended, versus generated by handwork which appears rough and uneven. Besides meeting print specifications, these last two will also create an aesthetic appearance noticeable to both experts and novices.

Your customer and the end user will frequently handle your components along with components machined by your competition. Attention to detail and improving blends and surface finishes will create an impression of superior quality. In the medical machining world this is extremely important. I received countless compliments from medical professionals and customers who could visually discern differences in appearance for both instruments and implants. Nearly all medical procedures require a "kit" of instruments. There is a high probability that the kit contains instruments from several manufacturers. Surgeons pride themselves on having a discerning eye and frequently will comment on small details. The orthopedic surgeons choose their implant supplier, which dictates the instrument supplier, as both are from the same manufacturer. The functionality and performance of the instrument kit are very important selection criteria, as the length of the surgery and the precision of the surgery are directly affected by the instruments. Since a manufacturer's sales representative is present at many of the scheduled implant procedures, the feedback on surgeon preference is well known. Orthopedic surgeons frequently consult on new products and test new designs. All orthopedic companies have Good Manufacturing Practices (GMPs) that govern edge breaks, blends, and machine lines. Surpassing these criteria in any industry is beneficial. Many companies break edges manually with deburr personnel. This is much easier for the CNC programmer and for the machinist, as they do not have to program and set the extra tooling required to break the edges and blend from one surface to another surface. Other companies will add the additional tools to the CNC program, add the tool paths to the program, and machine and blend edge breaks. I would argue that this can actually be a cost savings, as incorporating all features into your CNC machining process is more productive than handwork. The visual appearance between the two different methods can be substantial.

Producing part features with high CPk or Cr will be recognized by the technical personnel at your customers and the technical personnel among the end users. Your parts will assemble more easily, they are likely to yield superior performance and they will last longer. If and when they are measured and analyzed by your customer, your consistent achievement of providing parts to the mean will be displayed, appreciated, and respected. Ideally, your customer has shared or you have determined the features on each component that is Critical to Quality (CTQ). These are the features that when held closer to the mean will produce superior performance or longevity. You can also expect that if your company demonstrates noteworthy performance over your competition on CTQ features, your customer will be willing to pay extra for your components.

Producing part features with high CPk and Cr will reduce borderline quality scenarios and allow machinists more time to run multiple machines. Higher Sigma reduces both the cost of bad quality and the cost of good quality.

So what differentiates good in-process inspection practices from bad practices? First, the gages and instruments used on the shop floor for in-process inspection should be correlated to the gages and instruments used in the quality lab for final inspection. Likewise, final inspection gages at your company should be correlated with your customer's receiving inspection. This correlation seems obvious but is rarely practiced at low- or medium-volume operations. It is much easier in high-volume applications. While it is ideal to utilize identical gages or instruments on the shop floor, at final inspection, and at your customer's receiving inspection, it is not always feasible. Many components have features machined to size at separate machining sequences. The component is transformed from a chunk of raw material to a semi-finished and eventually a finished component during the course of 5, 10, or even 15 manufacturing operations. The component is a different physical size and shape throughout this process. This normally dictates that different types of instruments and methods are required to measure features "during" the manufacturing process, as opposed to the instruments and methods that are utilized at final inspection "after" the entire part is completed. In other words, the machinist can only measure the features of the parts that are already completed. Final inspection can measure the entire part. The machinist is dealing with a partial component that may not have datums and other surfaces finished. He has no choice but to measure differently and hence induce potential measurement variation.

While it is optimal to machine the entire part completely in one setup (one-and-done), this just is not feasible on many components. One-and-done enables identical instruments on the shop floor as in final inspection.

Machinists need to verify that each setup was correct by measuring the first piece. The machinist needs to measure subsequent components at some frequency to offset for tool wear and to determine when to replace tools. If the machinist's in-process measurements vary from the final inspection measurements, then you may be scrapping good parts and passing bad parts (see Figure 2.2). Likewise, your customer may be rejecting parts that final inspection has passed (#11, 12, 13). In Figure 2.2 it is obvious that you have scrapped components that your customer would have accepted. In this example part numbers 1 and 2 would not have been shipped even though the customer would have accepted.

	MEASUREMENT SHIFT					
	SOURCE (MACHINE)		FINAL INSPECTION		CUSTOMER	
PART #	MEASUREMENT *	RESULT	MEASUREMENT *	RESULT	MEASUREMENT *	RESULT
1	1.9982	GOOD	1.9978	SCRAP	1.9986	GOOD
2	1.9983	GOOD	1.9979	SCRAP	1.9987	GOOD
3	1.9984	GOOD	1.9980	GOOD	1.9988	GOOD
4	1.9988	GOOD	1.9984	GOOD	1.9992	GOOD
5	1.9991	GOOD	1.9987	GOOD	1.9995	GOOD
6	1.9994	GOOD	1.9990	GOOD	1.9998	GOOD
7	1.9998	GOOD	1.9994	GOOD	2.0002	GOOD
8	2.0003	GOOD	1.9999	GOOD	2.0007	GOOD
9	2.0012	GOOD	2.0008	GOOD	2.0016	GOOD
10	2.0016	GOOD	2.0012	GOOD	2.002	GOOD
11	2.0018	GOOD	2.0014	GOOD	2.0022	SCRAP
12	2.0019	GOOD	2.0015	GOOD	2.0023	SCRAP
13	2.0023	SCRAP	2.0019	GOOD	2.0027	SCRAP

* MEASUREMENT SHIFT DUE TO LACK OF CORRELATION

TOLERANCE 1.998" - 2.002"

Figure 2.2

The best machining organizations have an in-process inspection plan for each machining operation performed. The leading machining organizations also understand that it is more important to have an accurate inspection method than an accurate machine tool. It is better to have both, but a machine tool can be adjusted to create a good part, providing your inspection process is driving you to make good decisions. An accurate machine tool with an inaccurate inspection process will simply be adjusted to produce a bad part. Measurement error, whether reproducibility, repeatability, or interpretation, is the number-one driver of technical problems, scrap, and rework at most machining organizations. Many companies will not show this as the root cause because of lack of effective problem solving, lack of metrological knowledge, or lack or requisite breadth of machining principles. Interpretation includes understanding of the GD&T, datums, and the results of the measurement itself. Interpretation of the results can be further refined to include when to offset a tool, which direction to offset, which tool is the root cause, when to replace the tool, or whether there are other causations.

The in-process inspection plan will include all features to be measured, which inspection device to use for each feature, the tolerance of each feature, the frequency to check each feature, a balloon number or print zone/page for each feature, and a method to record the result of the measurement, who performed the measurement, and when it was performed.

We refer to machining steps as machining operations, which invites comparisons to operations performed by medical professionals. The operating room is run by the surgeon and the machining operation is run by the machinist. Both are dedicated to eliminating variation by utilizing the identical tools and instruments each time the same operation is performed. Each strives for the identical goals of high quality, predictable cycle time, no rework, no surprises, and certainly no scrap. The doctor tracks his patient's vitals through instruments for blood pressure, temperature, heart rate and more sophisticated equipment such as x-ray, CAT-scan, and MRI. A machinist provided with an inspection plan, tool change plan, the correct instruments, and magnification devices to review tool wear and the small changes to his parts also becomes capable of tracking his patient's vitals and diagnosing problems with his parts. The machinist will learn to connect tool wear with acoustical changes in the cutting process. He will learn to connect tool wear and acoustics to changes in surface finishes and slight dimensional changes. The machinist can prevent his patient from becoming sick, and if the patient becomes sick, he identifies the source quickly and cures his patient before surgery (rework) or death (scrap).

Providing the instruments to your machinist empowers him to be great. The machinist has a validated CNC program (which incorporates verification of workholding, fixtures, and tooling), an accurate machine tool, and a proven inspection methodology for each dimension. Virtually all variation has been eliminated with the exception of tool wear. When a dimension approaches a control limit, the machinist has already determined whether he will enter an offset to compensate for the wear or replace the tool if the total life has been consumed.

Diagnosing patient symptoms is simplified through elimination of lot-to-lot and time-to-time variation, because the operation is always set up with the same validated package of the CNC program, workholding, cutting tools, holders, and an in-process inspection plan.

Whenever practical you should utilize Statistical Process Control (SPC) and collect data electronically. If you are not utilizing SPC, it is still beneficial to collect data electronically. Devices increasingly provide more connectivity with lesser cost. There are many SPC software applications that simplify data collection, storage, graphing, and reporting. It is beneficial for machinists to view graphical trends and review stored data from the previous shift or previous year. SPC is an ideal form of in-process inspection that simplifies decision making while aiding the machinist to maintain dimensions centered in the tolerance zone according to Six Sigma methodology.

SPC is generally applied to higher-volume production runs. Even on these higher-volume production runs it is not feasible to utilize SPC on every feature or every dimension.

A newer approach to performing in-process inspection is the Coordinate Measuring Machine (CMM). The CMM has long been utilized for complex components in final inspection. A CMM has traditionally required environmental control for temperature, humidity, and cleanliness. Improvements in design and technology have enabled CMMs to be located on the shop floor. Not every environment is suited for a CMM, but many shops currently possess airconditioning and a level of cleanliness suitable for the equipment. If your shop environment is not suitable for a CMM, then the cost of a small, enclosed room adjacent to your machine tools for the CMM is not prohibitive.

A CMM can be programmed to measure most dimensions on each component for each operation. The normal process is for a member of the quality department to write, debug, and validate each program. The machinists can then load their components 24/7 on the CMM and start the program. The benefit of this method is that the CMM is exponentially faster than a human when measuring multiple features on the same part. Humans will have variability (reproducibility) from one human to the next and some variability within the same human. The CMM can transfer data to your SPC, and the CMM can quickly print a detailed summary of all measurements.

In addition to being fast and accurate, there are two tremendous advantages to your machining operations from using CMMs on the shop floor. First, the CMM eliminates the need for the machinist to spend time manually measuring components. This additional time directly contributes to your machinist being capable of not only running multiple machines but also of keeping the spindles more fully utilized. Even when the CMM is actually measuring a component, the machinist can step back and attend to his machine tools. Again, running multiple machines is the Holy Grail for machining productivity. Empirically, the machining cycle time to inspection time ratio was 8:1 or greater. With the advent of high-speed machining, combination machines, advanced cutting tools, etc. this ratio has been sinking fast. The time required by the machinist to inspect and record his dimensions reduces his ability to run multiple machines. Utilizing CMMs and other automated metrology moves the ratio back towards the traditional 8:1 and facilitates operating two machines.

Secondly, the CMM report summary can be printed with several user-friendly features that further reduce the machinist's involvement by preventing the necessity to record data. Since there may be dozens of features machined at each operation, this is an important time savings. More importantly, the printed report can highlight defined borderline measurements in yellow, and out-of-tolerance measurements in red. Since a machinist is ideally operating multiple machines, he may be required to view several pages of data to ascertain which tools need to be offset or replaced. This can be overwhelming. The color coding saves time and helps reduce BQ by insuring that features whose dimensions are not centered about the mean are properly addressed and corrected by the machinist.

Properly programmed, the CMM report will also display visual aids assisting the machinist to comprehend and respond to complex GD&T data. True position features can be displayed with a "crosshair target" (see Figure 2.3), and size features can be displayed with a bar chart. The most advantageous visual aid is the "profile" display, which will print the perfect "high" profile and the perfect "low" profile. The "actual" profile of the measured part will be displayed with the "high" and "low" boundaries—providing your profile is within tolerance. If your actual part profile is not within tolerance, it will overlap one or both of the boundaries (see Figure 2.4). This is powerful information, as it quickly articulates to the machinist, programmer, and inspector where in the profile is the problem or where in the profile is there a BQ situation. In this case, a picture speaks for a thousand numbers. Without this visual aid the problem solving and debugging would be much more extensive. This will save multiple technical people considerable effort and time.

Feature	NOMINAL	+TOL	-TOL	MEAS	DEV	OUTTOL	BONUS	
LOWER_RIG...	0.7343	0.0004	0.0004	0.7343	0.0000	0.0000	0.0004	[bar]
LR_TRP Position		IN			⊕ ⌀0.0004 Ⓜ A B C			
Feature	NOMINAL	+TOL	-TOL	MEAS	DEV	OUTTOL	BONUS	
LOWER_RIGHT	0.0000	0.0004		0.0000	0.0000	0.0000	0.0004	⊕

Figure 2.3

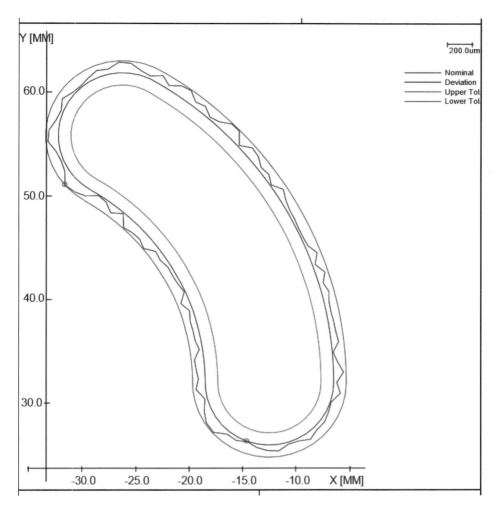

Figure 2.4 Photo Courtesy Zeiss Inc.

The combination of advanced CMM programming software, digital models, climate friendliness, and affordability have made the CMM an excellent choice for in-process inspection. I cannot stress enough the cost savings, quality improvements, and hidden benefits that utilizing a CMM on the shop floor will create. Connectivity of the CMM directly to the SPC or directly to the machine tool for offsetting is available and will become more sophisticated with Industrial Internet-of-Things (IIoT) advancements.

The availability of ample, real-time, and very accurate data is another excellent opportunity to combine digital technology, advance equipment, and people to create robust processes and problem-solving methodologies that turn your team into root-cause-correcting Ninjas on steroids.

GEOMETRIC DIMENSIONING & TOLERANCING

Geometric Dimensioning & Tolerancing (GD&T) is an international standard for communicating the geometry and tolerance for each feature on a print, electronic drawing, or model. If you are not knowledgeable of this standard, you need to educate yourself! If your workforce is not knowledgeable, they need to be educated. Why? GD&T is simply the language of machining, engineering, and metrology. These standards were created to eliminate the profuse notes that were written on prints to describe the relationship between different surfaces. The relationships include perpendicularity, parallelism, concentricity, etc. Notes are frequently vague, misinterpreted, and/or overlooked. The notes were written in the engineer's native tongue and were difficult to interpret. GD&T was also designed to clearly define the required functionality and permit additional tolerance if available. GD&T has been increasingly adapted since the early 1990s and is now widely applied in most industries. There are two primary GD&T standards, which are ASME Y14.5-2009 and ISO 1101. There are very small differences between these two standards when considering their full scope. Though, I expect there will be consultants who would differ.

GD&T is a language of symbols, numbers, and rules. It is to engineering and manufacturing as sign language is to the deaf and brail to the blind. If you have learned the system, a machinist in Japan and in the U.S. will both interpret a print created by a German engineer in an identical manner. Receiving inspection in France will inspect the part with no translation, and the service personnel in Australia will use the print for problem solving.

GD&T is not a standard that can be taught in a few hours. The training is a minimum of 24 classroom hours. An individual will only become skilled with extended application and constant referral to a handbook. There are also a number of self-paced tutorials that teach employees print reading, basic GD&T, and advanced GD&T principles. The best tutorials provide examinations at the end of each section. An employee who does not pass the examination will repeat the section until he/she can pass the test. Employees can be scheduled one hour per day or one hour every other day until they complete the program. This type of schedule is more effective than placing someone in a classroom for eight hours who is not conditioned to sitting and consuming detailed information for lengthy periods of time. The GD&T tutorial will also be available for new employees.

How important is GD&T? Anyone who is quoting new work must be highly skilled with GD&T to competently determine the process, hours, and cost. Two widgets that appear to be physically the same to the naked eye may indeed be two completely different animals. Widget A was created by a design engineer who understood the functionality of his component and toleranced the print with the proper use of GD&T. Widget B was created by a design engineer who sent his model to an intern for detailing. The intern did not understand the functional requirements but wanted to be cautious to prevent receiving any blame if Widget B might fail, so he added a few additional feature control frames. Widget B now became twice the cost of Widget A, as the difference in GD&T added two superfluous grinding operations.

All of your technical personnel must be competent with GD&T. This includes engineering, programming, machining, and inspection. They simply cannot perform their jobs effectively without this working knowledge and application of GD&T. Additionally, your sales team needs a basic understanding, as they will not be able to comprehend the value of what they are selling without this knowledge. Operations management personnel are no different. They cannot understand basic problems and make good decisions without speaking the language.

GD&T will be discussed in several other chapters of this book, as its tentacles reach into significant areas of the machining business. I am not suggesting that your employees learn Latin. GD&T is equivalent to Microsoft Office, basic math, or even reading. It is a necessity wherever you may reside on the organizational chart.

RISK ANALYSIS

Sometimes it is more important when you say "no" than when you say "yes". Whether you are an OEM or a contract manufacturer you need a formal process to review the risk associated with a new project. Risk can take several forms. The highest risk is a quality risk. If you accept an order and later determine that you cannot adhere to the specifications, you will need to notify the customer. If you are an OEM, you may also be the customer. Your peers in design or marketing can decide on plan B. You may or may not need to discuss with the end customer. If you are a contract manufacturer, your customer may not be so understanding. Your customer may demand that you adhere to all tolerances and specifications regardless of time or cost. The customer may force you to outsource a portion of the project or they may cancel the order and award the project to your competitor. Regardless, the outcome will be some degree of damage to the relationship with the customer and certainly a high degree of damage to profits.

Another form of risk is delivery. If you do not properly understand the specifications and the magnitude of the project, your delivery has a high potential to significantly surpass the negotiated delivery date. The ramifications will be once again damaged relationships with accompanying cost overruns. When deliveries are late on critical projects, it is a normal result to throw money at the problem in terms of overtime, expediting, and additional resources. In fact, cost is our third form of risk. A sizeable monetary loss is created not only by quality and delivery entanglements, but also by the failure to understand the full scope of the project.

The root causes of the risk on new projects are frequently, but not exclusively, new internal or external processes, tight tolerances, size (large or small), new materials, GD&T, or a combination of these factors. Design of Experiments (DOE) and problem solving experts will reinforce that the last scenario, a combination of factors, is the most challenging to predict and diagnose. The sales folks normally see dollar signs and not warning signs. There are some sales professionals with enough technical background to spot the risk associated with a new project, and many that do not possess an adequate technical background to identify risk. Whoever performs the quoting needs to summarize any potential risk and conduct an appropriate design review. Depending on the scope of the project this design review may be an informal discussion or an in-depth cross-functional review. This is the time to identify request for changes to the print and send the feedback as Design for Manufacturability (DFM) advice. During this process it is common to find various print errors and omissions. It is also common to find GD&T mistakes or GD&T requirements that significantly add cost to the project. In many of these instances it is sufficient to submit a conditional quote based on the acceptance of your requested changes. I have been awarded major projects simply because the customer's engineering group recognized my organization's level of comprehension of the details. Customers will frequently award projects to those suppliers who have reviewed the specifications with a great deal of thoroughness. They comprehend that these suppliers are more likely to deliver on time and with higher quality.

I have occasionally produced sample parts to evaluate risk prior to quoting or committing to the project. There are occasions when lower-cost test pieces can be utilized to simulate machining, heat-treat deformation/growth, coating build-up, etc. If you really need to know how a component will move after significant stock removal, it is feasible to simulate only that portion of the process rather than produce the entire component and thus determine your answer more quickly and more economically.

I am not aware of studies that have focused on manufacturing companies that have closed to determine the root cause(s) of their failure. I expect that accepting orders that did not match their capabilities would be high on the list. If you ask "why" enough times, you will find poor decisions and negative events traced back to accepting projects without the requisite skills and equipment. Sometimes this is one or two large projects, and sometimes it is a pattern of smaller orders.

I led a turnaround of an orthopedic instrument and implant acquisition that had previously underestimated the cost and quality risk of a high-dollar instrument project. The organizational veterans regularly referred to the project as the turning point in the organization from success to trouble. Fortunately, as is the case with many medical projects, the initial release quantities were extremely high, but the subsequent orders were low-volume replacements by the time I was involved. With several years of improvements and corresponding price increases the project eventually became break-even.

In the end this comes back to leadership and strategic vision. If a project involves risk, with little or no business sense, it needs to be rejected. If a project involves risk but advances your strategic vision by acquiring a new customer, adding a niche or skill offering to your portfolio, or simply through significant revenue opportunities, then the risk needs to be solved and converted to a strength.

Your quoting and production/design review process needs to be sophisticated enough to proceed with non-risk projects and yet sophisticated enough to also elevate legitimate risk for further analysis. Your leader needs to have the technical and business skills to make the right decisions. More importantly, he needs the fortitude to say "yes" to an operations group wary of the challenge and "no" to a sales group looking for commission. It is not uncommon for sales to be lobbying for quick acceptance of a new project, while operations is insisting that the price, the delivery, or the quality requirements—or all three—are not achievable. There are frequently dominant personalities and high emotions involved. Sales will argue that they have been working for this opportunity for years, it will be years before a similar project comes along, and of course that "the competition can deliver the project so why can't we". Operations will counter with "what current projects will be moved back", "we can't measure these features", and "we will be sending a hundred-dollar bill with every part".

Yes, it is up to your leader or your leadership to navigate the technical and commercial issues associated with projects that challenge your capabilities as an organization. Whatever technical hurdles that cannot be reasonably resolved prior to order acceptance may be mitigated through negotiation and contractual language. When this occurs, someone has to take charge. There is a great deal at stake in such situations.

ROOT CAUSE AND CORRECTIVE ACTION

In a manufacturing environment, failures occur all the time. The best companies and the best cultures own their failures, track their failures, and reduce their failures. Mistakes, errors, screw-ups, accidents, faux-pas, rework, scrap—they all occur 168 hours per week in a fast-paced facility.

You are not going to conduct a Kaizen event every time you find a hole out of position or a scratch on a part. After you summarize and evaluate the non-conformance data you may determine to conduct a Kaizen event to eliminate chronic failures. The larger question becomes "how" to collect the data and "what" events to collect data about.

It is common to enter a Non-Conformance Material Report (NCMR) into a data base. There are many third-party software packages that collect all relevant data, send email notifications, and provide summary and detail reports. These packages facilitate administration and compliance of the NCMR process. Most also provide entry and tracking of preventive actions and continuous improvement opportunities.

Most importantly, an NCMR package provides a database to track quality issues by part number, workcenter, machine, employee, shift, or part family. If there is a problem in the middle of the night, there should be an NCMR waiting in the inbox of the appropriate person in the morning when he/she arrives. The individual needs to disposition the non-conformance and determine whether the root cause of the failure has been corrected. This is accomplished by incorporating a Corrective and Preventive Action (CAPA) section into every NCMR to insure that a root-cause corrective action is implemented. When problems are fixed immediately, they do not escalate. It is common in the medical industry to receive new orders for implants or instruments for multiple size ranges and for both left and right hands. These types of part families are prevalent in many industries. When production is ramping, it is a necessity to receive immediate feedback on any complications. There are likely many more fixtures and programs to be created that should incorporate the root-cause corrective action. This alone is a huge cost savings. Even for long-established products it is critical for engineering, programming, and quality personnel to see the details behind shop problems in a timely manner. Everything learned during the corrective action can be applied to existing and future components that contain similar features or processes.

It is important to be a learning organization. To become a learning organization, you need to have learning individuals. To foster learning individuals, you must provide timely data in a format they can absorb, act upon and disseminate. You need to have expectations and a culture to get things done with a sense of urgency and thoroughness. Everyone gets interrupted, everyone has a few balls in the air— just get it done!

Many manufacturing organizations process new parts every day. Many organizations manufacture parts that they have not made in months or years. This type of activity requires constant problem solving and refinement of fixtures, CNC programs, tooling, inspection gages, etc. For those problems that are not discovered immediately and/or cannot be fixed immediately, a closed-loop NCMR process insures that the individual(s) responsible, including outside suppliers, receive timely information and are held accountable for a corrective action to the root cause.

Let us look at a few of the benefits of a good NCMR database:

- Generation of in-depth metrics that not only speak to the magnitude of quality issues but paint a picture of who, what, when, and where.
- Justification for new equipment, accessories, or processes.
- A trove of quoting support. If repeat work is showing low margins or negative margins, the NCMR database will display all quality-related problems and provide confirmation that root causes have been implemented. Instead of raising the price and losing the order, perhaps the price should remain constant.

FINAL INSPECTION

Final inspection is sometimes referred to as the quality lab, metrology, QC, or simply inspection. Each facility generally has one or more locations where much of the inspection equipment and inspection personnel are located. It can be beneficial for parts or assemblies to be inspected or simply approved for use in a cellular environment versus final inspection. Even in these cases the final inspection area will support and audit the cells.

Final inspection is generally involved in or responsible for the following:

- Receiving inspection
- First articles from suppliers and first articles going to customers
- PPAPs (Pre-Production Approval Process)
- Calibration of instruments and gages
- Gage R&Rs
- Verification of internal setups, whether considered in-process verification or a first article
- CMM programming
- Designated Quality Representatives (DQR, customer dock-to-stock program)
- Approval prior to shipment—may or may not be via formal sampling plan

Machining environments vary, but the final inspection department is normally a beehive of activity. The list above contains numerous tasks that are critical to suppliers, internal operations, ISO compliance, customers, and profit. All of these items are technical tasks that will be audited and will have lasting ramifications. These tasks also must be completed on a timely basis, or projects will be delayed and equipment will be idle. Most organizations employ a combination of quality engineers and inspectors to complete these tasks. A normal ratio of inspectors to quality engineers and quality managers is about 5:1, depending on the industry.

Mistakes and errors in the final inspection department generally lead to expensive and embarrassing predicaments. This group is required to correlate measurements between the supply chain, operations, final inspections, and the customer. This group is also expected to be the second pair of eyes on all decisions made by engineering, programming, or operations. This means that they should not assume that anyone's calculations, interpretations, or judgments are correct. The final inspection department should only use the final print and any specifications either on the drawing or referred to on the drawing.

Better machining companies have better people, better equipment, and better culture in their quality lab or final inspection department. It is an obvious recommendation to move talented people to this area, who already possess a wealth of technical knowledge, provide training, encourage education, and obtain metrology equipment with the highest accuracy that is suitable for your families of components.

The last word on the quality lab and final inspection department is that this area is much more productive and effective when good parts are produced. When non-conformances occur, then 100% inspection will have to be performed. Worse, the 100% inspection may be required multiple times on the same order. This is an obvious statement, but the effectiveness to create robust processes and provide quality at the source is poor at many machining organizations.

METRICS

A discussion of metrics could occur in almost any section of the Machining Pyramid. I have chosen the quality system building block, as much of the data for corporate metrics is created through metrology and the NCMR database. Additionally, the ISO procedures generally require the metrics to be reviewed during the quality management reviews, which are coordinated and led by the head of the quality organization.

It is well understood that "we get what we measure". The data-driven digital world creates ample opportunities to create dozens of metrics. It is more likely that organizations suffer from data and metric overload than data and metric shortages. The purposes of the metrics are:

- to understand the performance of people, departments, and facilities;
- to provide data and information, so that all people can take responsibility and action to fix problems;
- to provide data and direction for management decision;
- to incentivize and motivate people.

Metrics frequently do not collect the data intended or targeted. The data is frequently not tabulated, formulated, and reported as intended. And finally, the reported data is frequently not interpreted correctly. For these reasons, it is important to validate or authenticate your metrics. Some of the most common operational metrics are listed in Figure 2.5. This does not include financial metrics and is by no means a comprehensive list. There are many creative and applicable measurements derived by organizations. It will behoove the organization, if the metrics the workforce is asked to minimize or maximize actually encourage beneficial behavior. Consistent behavior does eventually become the culture.

COMMON MACHINING METRICS

SAFETY	DELIVERY	QUALITY	PROFIT	LABOR	SUPPLY CHAIN	INVENTORY	OTHER
DAYS SINCE LAST ACCIDENT	ON-TIME BY DOLLARS	PPM DEFECTS	GROSS MARGIN	UTILIZATION & EFFICIENCY	SUPPLIER ON-TIME	INVENTORY TURNS	PERISHABLE TOOLING/HOUR
# DAYS LOST TIME	ON-TIME BY PIECES	SCRAP/REWORK TOTAL DOLLARS	MARGIN PER FAMILY	INDIRECT HOURS PER CODE	PERCENT DEFECTS	WIP	PERISHABLE TOOLING % VALUE ADD
# INCIDENTS	ON-TIME BY LINE ITEM	SCRAP/REWORK % VALUE ADD	PROFIT FROM NEW PRODUCTS	PLANT EMPLOYEE UTILIZATION	SUPPLIER CYCLE TIME	INVENTORY DOLLARS	
# NEAR MISSES	AVERAGE DAYS LATE	TOTAL COST OF QUALITY	PROFIT PER EMPLOYEE			INVENTORY ACCURACY	
INJURIES PER 100,000 HOURS		NCMR PER ORDER	RETURN ON EQUITY			OBSOLETE $	
		NCMR'S PER MONTH	RETURN ON ASSETS				
		% NCMR'S FINAL INSPECTION	NET PROFIT MARGIN				

Figure 2.5

3 PEOPLE

All companies have access to the same equipment. All companies have access to the same software. All companies have access to the same tooling, workholding, gages, etc. So what makes the difference between the winners and the losers, the high-profit organizations and the break-even organizations, the growing and going out of business? The answer is: your people! It is your people who determine which software, which machine, which tooling, and which orders to accept or reject. It is people who innovate, people who create robust processes, and people who develop customer bonds that outlast ownership changes, management changes, mergers, and delivery or quality emergencies.

At some point, every organization will be faced with a crisis of their own making. This may be a large warranty claim from sins of the past, a recent quality escape jeopardizing a customer's reputation, or a severe delay in delivery, causing your customer's high-profile project to miss release dates. During these emergencies it is your people who will help plot the recovery, and it is your people who will execute the recovery plan. Equally significant, it is your people's deep-seated relationships they have nourished with the customer that will ultimately prevail.

Instead of using the term associate, employee or team member, I prefer to simply use the term "people". I am a person, you are a person, our friends, neighbors, and family are each a person, and collectively we are people. There are some contexts in writing in which the word "employee" is necessary to communicate properly, but in this book you will understand the subtle difference. We do not stop being a person while at work. Everyone at your organization is a person. People have pride, integrity, intelligence, motivation, and of course, emotions. People also possess the crucial creativity and innovation. These human qualities are what separate great organizations from poor companies. Since none of us have discovered how to segregate pride, integrity, and motivation from human emotion, we necessarily must embrace all facets of human behavior in our leadership and decision making.

When people are treated like people, the long-term results are far superior. Individuals with talent and skill can choose where to work. They can work for you and attract other people with talent and skill, or they can go somewhere else and attract the rest of your people with talent and skill. I call this "A-players attracting other A-players". Why is it so important for machining companies to have "A-players"? The pace of change, increase in technology, and global competition requires more innovation to achieve the productivity and quality levels. Machining historically was a skill- and knowledge-dependent profession, but the changing market dictates even more emphasis on talent at all levels of the company.

Personally, when I write, read, hear, or speak the words "person" or "people", I make wiser decisions. I am forced to be reminded that the topic requires consideration to perceptions, emotions, culture, and individual traits. Unlike a formula, machine, or spreadsheet, when the topic is people, the ability to project the outcome will require more depth and consideration. The outcome will be far less predictable. My decision will not look for a black or white outcome, a win or lose outcome, or a short-term outcome.

A leader wants a higher contribution from all of his individuals, regardless of talent level, stage of career, or job title. Sure, there will be a few people who will not respond to any leadership and will need to be addressed. Some organizations change policies and punish everyone due to the actions of a few or one individual. I have found that it is much more effective and respected to address and correct the problem at the source—with the one individual, rather than change policies that negatively affect the many.

Good leadership does not really "get" more from employees over the long term. People will "give" more to good leadership.

The first step is the recognition at all levels of the organization that identifying, hiring, developing, and promoting good people is a strategic necessity and paramount to success. There are many managers in key positions who believe that if you buy the newest and fastest machine tool, you will beat the competition. Possessing good equipment is an important building block to our Machining Pyramid, but unfortunately even the best machines do not run themselves. They cannot perform their own maintenance, write their own CNC programs, change their own tools, or perform their own inspections. Quite frankly, the newest and fastest machine tool with a few wrecks, or poor maintenance, or lack of optimal programming/engineering will be no different than any other horse in your barn. A good team of people with good execution will achieve a ROI significantly faster than an average or poor team of employees. Good people can do more with average equipment than average people can do with good equipment. However, optimal productivity requires the best of all.

What is the definition of a good employee or a productive person as a machinist? Achieving productivity is a combination of several factors. I have created what I consider to be a productivity multiplier (see Figure 3.1). Skill is mandatory, but machining skill can be better described as knowledge & innovation (KI). Some machinists are skilled but not productive, so motivation & effort (ME) are mandatory. Machining productivity cannot be separated from the level of investment or the level of productive capital (C) equipment provided to the machinist. I will describe in later chapters why productivity is a multiplier and not just incremental or additive.

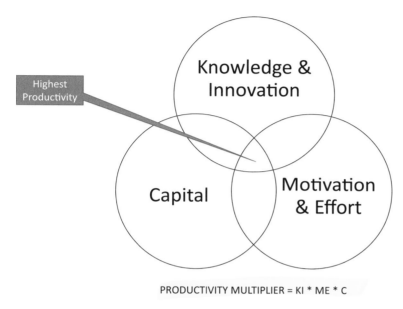

PRODUCTIVITY MULTIPLIER = KI * ME * C

Figure 3.1 Productivity Multiplier

For now, it is important to recognize the productivity multiplier: KI X ME X C

It is true that some high-volume operations can be successful with a lower talent level of machinists, engineers, and programmers. This is possible by buying equipment in a "turn-key" manner from the manufacturer. In general, if your equipment runs the same open tolerance components for weeks, months, or possibly years, then you are likely to employ operators and not machinists. The difference is that a machinist can set up the machine, debug programs, inspect parts, and troubleshoot problems. An operator will generally load parts, replace tools, and perform simple inspections.

There is nothing wrong with employing operators. Many companies have a mix of operators and machinists. It should be one of your team's goals to simplify as many jobs as possible, so you do not need a high-skilled machinist everywhere. This helps the nightshifts and weekends run more smoothly. It helps new employees, whether they are operators or machinists, to contribute in a shorter period of time. There are many traditional approaches to standardize and simplify the process. Newer techniques include closed-loop and Adaptive Machining, which utilize scanning and probing to reduce the human time and skill necessary to produce complex components.

It actually helps the machinist earn higher wages, as employing some lower-paid operators creates payroll space to pay premium wages to machinists. Understanding the skill level for different facilities, departments, or equipment is important. Quite frequently, an employee new to the machining trade will be an operator for a number of years, but with motivation and intellect will eventually develop the skill to become a machinist whether with his current employer or with a new employer. Many operators do not have the motivation or ability to become machinists. For all these reasons it is beneficial to employ operators. However, many jobs simply require the skill of a machinist. If you need to determine where your organization ranks on the machinist vs. operator scale, you should complete the following exercise to determine your company-specific skills and hence compensation level.

MACHINISTS' SKILL LEVEL

You must rate your products in the following three categories: tolerance, material type, and volume.

Tolerance determines the accuracy and maintenance level required of your machine tool. It determines the type of measuring device, cutting tools, workholding, etc. More importantly, the tolerance determines how closely your personnel must inspect the components, offset the tools, and change the tools. Maintaining close tolerance print dimensions effects every type of machining, whether turning, grinding, EDM, honing, or any unconventional machining process such as electro-chemical. Rate yourself on a scale of one through ten, with ten as the most difficult. Location tolerance above .030" is a one, and location tolerance below .0002" is a ten (see Figure 3.2). Location tolerances are more difficult to achieve than size tolerances. Size tolerance above .015" is a one, and size tolerance below .0002" is a ten. Complex profiles are more difficult than diameters and slots. If a large part and a small part have the same tolerance, the large part will be more difficult to machine to print specifications. Application of Geometric Dimensioning & Tolerancing (GD&T) may make some components easier and others more difficult. Overall, you should be able to directionally determine whether your tolerance difficulty is a one, a five, a ten, or somewhere in between.

Next, you need to rate your material hardness one through ten. Material hardness determines how fast your cutting tool will wear. Tooling that is wearing quickly will necessitate constant part inspection, regular adjustment to length and radii offsets between tool changes, and actual replacement of the cutting tool and re-setting of the offsets.

Hard material will also more readily create cutting tool deflection and limit depth to diameter ratios. All this requires much more attention and skill from the ma-

MACHINIST SKILL

TOLERANCE (inch)			MATERIAL		VOLUME	
LOCATION	SIZE	SCORE	TYPE	SCORE	QUANTITY	SCORE
±.030	±.015 DIA	1	PLASTICS	1	>5000	1
±.020	±.010 DIA	2	COPPER/ALUMINUM	2	>2000	2
±.015	±.005 DIA	3	LOW CARBON	3	>1000	3
±.010	±.010 PROFILE	4	TITANIUM	4	500 - 1000	4
±.005	±.005 PROFILE	5	CASTINGS/WELDMENTS	5	100- 500	5
±.002	±.001 DIA	6	STAINLESS	6	<100	6
±.001	±.0005 DIA	7	TOOL STEELS	7	<25	7
±.0005	±.0002 DIA	8	WASPALLOY C	8	<10	8
±.0002	±.002 PROFILE	9	INCONEL	9	<5	9
<.0002	<.002 PROFILE, <.0002 DIA	10	COBALT CHROME	10	1	10

MACHINIST SKILL INDEX = TOLERANCE + MATERIAL + VOLUME

Figure 3.2

s. The machinability rating for a basic aluminum alloy such as 6061 is 1.9, he machinability rating for Inconel is .2, with the lower number more diffi- machine. If you machine soft materials such as aluminum, your rating should be a one or two, while materials such as Inconel or cobalt-chrome will be a nine or ten. Other materials will fall somewhere in between. The exception is for castings and weldments. Due to the inconsistencies in the geometry of these items increase your ranking by one level from the recommendation in Table 4.2. This may not contribute to machinability in the same manner, but it has the same effect of increasing the difficulty required to machine the components to the print.

The final skill level factor is your volume. Some companies are low volume/high variety, while others are high volume/low variety. Low volume dictates that your equipment will regularly need new setups or changeovers. The setup frequency may be several times per day, once per day, or once every few days. High volume allows for much fewer setups on each machine. Whether your frequency is once per week or less, a higher-volume shop with dedicated equipment reduces the overall skill level. Minor changes to a machine or process to switch to another component in the same family do not constitute a setup or changeover. An example is a Swiss-style lathe switching to a longer or shorter component.

Frequent setups require a higher number of overall employees with greater ability. A corollary to more setups is more new CNC programs, fixtures, tooling, and inspections. Besides increasing the skill level requirements, more setups dramatically increase the complexity to manage and schedule your facility. Rate yourself a one, if your machine runs for a month without a changeover, and a ten, if you change over once per day.

Now apply the formula:

TOLERANCE + MATERIAL HARDNESS + VOLUME = SKILL LEVEL

If your skill level is less than 8, you generally have operators with some setup machinists for the occasional changeover, new part, or training of new employees. A score between 8 and 15 will require a higher ratio of machinists than operators. If your score is over 15, you will need a majority of high-skilled machinists. As your score climbs significantly above 15, the escalating amount of training and skill required for the workforce becomes obvious.

It is common for many machining organizations to contain some departments and/or part families that will score very high on the skill level, and other departments and/or part families that will score low. It is beneficial to apply this exercise to your different departments and part families, as it will determine where you may need separate compensation and skill levels.

TECHNICAL POSITIONS

As we all know, a manufacturing organization requires many additional skilled support positions. These are the people that must create, implement, and execute the innovation required to beat the competition, and they are the team that must stay current with the rapidly changing technology. The universal technical positions required at a machining organization are a CNC programmer, a manufacturing engineer, a CMM programmer, a quality engineer, and a supervisor. I will discuss each one individually:

CNC PROGRAMMER: Since your company will spend 150k, 500k, or more on a machine tool, you will need someone talented to write the programs. Let us not forget what the acronym CNC means—computer numerically controlled. A human has to create these programs. A wrecked machine, an idle machine, or an inefficient program will not make the company money but cost the company money. Simulation software will continue to play a larger role for the efficiency of the CNC programming process. Effectiveness of the simulation is a byproduct of the CNC programmer developing accurate kinematics of the machining environment and presenting the appropriate fixtures, holders, and tooling to the simulation process.

There is a very limited number of people who are very good at CNC programming. CNC programming requires a thorough knowledge of machining fundamentals, workholding, cutting tools, tool holders, metrology, and the CAM software that your company employs. Solid models assist the process, but a CNC programmer must have a mind that conceptualizes complex three-dimensional shapes. He must possess good mathematical skills and be very analytical to resolve the smallest of details. In addition to these skills, the CNC programmer needs to be a strong communicator to incorporate the ideas of the machinists and the engineers into each project. I refer to the creation of the CNC program as a project, since each program needs to be accompanied by setup documentation, tool lists, and possibly special fixturing and tooling. Each of these will need to be debugged along with the hundreds and possibly thousands of lines of programming code. Like any team dynamic, incorporating input from others creates a more successful project.

To find all these skills united in one individual is rare. There are two backgrounds that seem to generate successful CNC programmers. The successful CNC programmer can be a highly skilled machinist who possesses enough writing skills to communicate with peers/customers/suppliers along with technical writing to properly create instructions to accompany his programs. More importantly, he needs to have the temperament to sit in front of a programming station for long periods, and the people skills to not kill his old friends who are now criticizing his programs.

The second background that generates strong CNC programmers is a degree in engineering in combination with a hands-on mentality. I have assigned some of these individuals to train with a top machinist and to spend time operating the machine tool while ideally debugging and running his own CNC programs. The engineering degree insures the mathematical skills, the analytical mind, and a determination to master the software. You have to help him obtain the machining experience to truly understand the nuances of surface finish, rigidity, tolerance, climb cutting vs. conventional cutting, etc. This training also allows the engineer to build a working relationship with the machinists and to earn some level of credibility.

Regardless of the background, it will take about a year for a talented individual to learn your CAM package and then to become a productive CNC programmer. It will be three to five years before he begins to reach his full potential. This timeframe may be more, if you are cutting complex shapes in more than three axes.

Besides creating and debugging your new CNC programs, your programmer will also be an important resource for solving technical problems, quality problems, and machine control issues. For all these reasons it is imperative to develop a stable of talented programmers and begin the search/training process early, when facing growth or attrition. It is wiser to have one too many programmers than one too few.

MANUFACTURING ENGINEER: The manufacturing engineer typically is responsible for quoting, make/buy, routing, equipment acquisition, and fixture design. It is critical to comprehend that your manufacturing engineer, along with input from the rest of your team, is responsible for developing a better mousetrap. This means that they need to create and refine a process for a component or family of components that is superior to any of your competition. This superiority can be achieved in many ways, be it by consuming less material or less machining time, by producing better quality, or thanks to more unattended machining, the process needs to provide a competitive advantage from your competition. This process development sometimes occurs before or simultaneously with the application of lean techniques and Six Sigma techniques. We will discuss this further in our review of the building blocks of our Machining Pyramid.

CMM PROGRAMMER: The Coordinate Measuring Machine (CMM) has become the dominate measuring device in the machining environment, and this trend will continue. The fundamentals behind this popularity is the increased use of GD&T, digital models, and the improved affordability of the equipment. The reasons are discussed further in the Quality Systems chapter, but at this time the result is a greater emphasis on the individuals who must write the CMM programs.

The background for CMM programmers varies between quality engineers and veteran inspectors. The individual must learn the chosen CMM programming language, which is normally provided by the CMM manufacturer. The CMM programmer must manipulate the solid model, decipher datum structures, determine setups, and translate GD&T requirements into CMM programs. The CMM programmer needs to possess ample metrology experience to validate the results of his/her new program. Since components are generally machined in multiple operations, the CMM programmer must create a setup and program for each machining operation plus a final inspection program.

Large new projects and smaller programs requiring velocity will inevitably require the CMM programmer to be in sync with operations. If the machining operations are ready, the metrology must be ready, or machines and people on multiple shifts will be idle. In general, high volume manufacturers will need a significantly lower quantity of CMM programs, while the low-volume/high-mix companies will require a large number of CMM programs.

QUALITY ENGINEER: We discuss in subsequent chapters that quality is also productivity. In the tight-toleranced machining environment you cannot separate metrology from the machining process. You must be able to create quality at the source—which is the machining operation—and correlate all measurements from the supply chain through to the end customer. Whether you employ multiple quality engineers or combine the position with a quality manager, this position is responsible for supply chain quality, internal gaging and instruments, and is the quality liaison with the customers.

SUPERVISOR: As with all people management positions, the machining supervisor requires communicational, administrational, and motivational skills. Unlike many managerial positions, the front-line machining supervisor requires strong technical problem-solving ability. Rarely will an individual possess the breadth and depth of technical skills accompanied by the managerial skills. Solving problems to insure quality, safety, and productivity normally is the higher priority, and hence machining supervisors tend to be technically inclined. Weekends and nightshifts have little or no support from engineering and programming, so the supervisor is critical for maintaining quality and delivery.

Given that the supervisor is involved with technical problem solving, the ratio of supervision to employees needs to be lower for machining than for other manufacturing.

4 LEADERSHIP

There are many books and seminars on leadership. I am only going to discuss leadership as it relates to machining success, growth, and profits. This is my area of expertise. What you know, what you have learned, what you were born with, and what you have developed about leadership is all applicable to machining organizations. Specifically, there are unique aspects of leadership, as it relates to a highly technical, dynamic, and people-oriented business such as machining. This is an industry where a strong education, being articulate, and possessing solid business skills are not enough. The leader must also be fundamentally sound in general engineering, quality, and machining principles. This is required to lead or participate in machine tool selections, interviews, acquisitions, expansions, new business opportunities, promotions, etc. I do not believe that you can run a legal firm without a working knowledge of the legal system, and I do not believe that you can run a software company without knowledge of the software business. The same logic is applicable to machining organizations.

A large enterprise may have a president who is an effective generalist. However, each machining organization or facility, regardless of size, needs a manufacturing or machining champion. This is the person that ultimately drives success and profits. Without this effective champion, capital purchases will not be optimal, A-players will drift away, growth will stagnate, and margins will be squeezed. In today's connected world, our customers can obtain quotes from Asia as simply as from North America. We are competing with organizations in low-cost countries across the globe as well as low-overhead competitors in North America and Europe. Your competitors can leapfrog your competitive position, and your competitors are also practicing continuous improvement. You have to improve at a faster rate than your competition, so you better have a leader who possesses all the tools to move your organization forward in all aspects of our Machining Pyramid.

Some specific initiatives a manufacturing and machining leader needs to provide:

- A great machining organization is about accumulating talent. A leader needs to hire and develop technical people at all levels of the organization. He simply needs the combination of emotional intelligence and technical understanding to attract and mentor A-players. Each company has a responsibility to develop professionals and leaders for tomorrow.

- Machining is a business that is affected by breakthroughs in material sciences, software, processors, metrology, connectivity, simulation, and much more. The leader needs to grasp how his business and his customers' requirements are affected by these changes and incorporate this comprehension into a vision that prepares his organization for success in the new environment and in the future environment.

- The leader needs to define and solve the problem or emergency currently at hand. We cannot always work on our strengths or according to a schedule.
  - Building success in an organization is a long process. The decline of the organization can be very swift. This fact is more germane to companies without an installed customer base requiring spare parts and without brand loyalty, i.e., an OEM versus a contract machining organization.
  - Losing key people, purchasing the wrong machine tools, a major quality escape, losing a key piece of equipment, or a vital customer leaving will all contribute to hard times.

- You must try to grow, or you will shrink!
  - Customers leave, customers disappear, customers shrink, customers get bought, product lines are re-engineered or replaced.
  - The leader needs to ask and determine why he is losing orders he would like to book, what will be the needs of his customers in 3–5 years and how he will position his organization to meet those future needs.

- The leader should understand his organizational niche in the market, improve this niche, move deeper into current customers and broader with other customers.
  - The niche can be strategically enhanced through new designs, offering DFM, better quality, SPC, better inspection criteria, better aesthetics, better CPk, design help, VMI, stock raw materials, R&D of future designs, etc.
  - The niche can be utilized to attract new customers, or the niche can broaden in the sense that the additional product/service offerings are tangential and can be leveraged off the niche.
  - The leader needs to have a broad understanding of the marketplace to understand how to prevent his niche from being moved to a low-cost region. This may mean offering such high quality, reliability or service that customers do not initiate a search or decide to invest their search effort on re-sourcing another product line produced by another supplier.

- The leader needs to have an impact on all the indirect personnel. Through better prioritization, motivation, communication, and a hands-on approach he should be able to make each person 5 % more effective—minimum. For every 20 indirect people this is like adding one more highly talented employee at no cost. If the leader achieves a 10 % effectivity increase, it is two additional people per 20 indirect. More importantly, it is a lot of projects and continuous improvement that otherwise would not be completed.

- The leader needs to be a builder, not a divider.
  - Teamwork is more effective than politics, and esprit de corps triumphs over self-promotion.
  - The leader establishes the culture of what type of personality and behavior equates to success, and what type of personality and behavior should move down the street.
- Of course, we want our leaders to be passionate about manufacturing, humble, optimistic, competitive, and willing to spread credit and praise to others when initiatives are successful. A touch of conversational intelligence and an understanding of team dynamics complete the package.

Conversely, there are leadership traits that should be red flags to owners or the board. Instead of developing, mentoring, and empowering people, your leader saps self-esteem through micro-management. Leaders who are copied on copious emails to make decisions on issues three levels below on the organizational chart not only lose track of the strategic goals but create a company where people are stagnant with paralysis, waiting for the leader to respond with the magic answer like Oz behind the current. Leaders more concerned with the amount of hours worked and transactional exactitude than actual results and innovation are incapable of talent development and sustained growth and success in a dynamic competitive market.

FAST-TRACK LEADERSHIP

I have experienced three approaches that organizations apply to attract and/or develop their leaders and champions. The first is the large company approach where high GPA graduates from prestigious universities are recruited into fast-track programs. These individuals are immediately placed into management positions. I hesitate to refer to these assignments as "leadership" positions, because the individuals do not have the experience to lead. They may have the potential to lead in the future, but their knowledge of the business and respect from their peers has not been established. They are rotated to new assignments frequently at different facilities and different areas of the business. Depending on the organization, the assignment typically lasts between six months and two years.

While it is beneficial to recruit high-potential employees, the shortcoming of this approach is primarily the duration of the assignment. This is compounded by subsequent assignments not being in the same facility, division, geographical region, or possibly the same business discipline. The prospective leader rarely gets to follow medium to large decisions and projects to completion and beyond. The development of an individual to his or her full potential is reduced when the individual is not present to see the results (good or bad) of his/her choices and actions. How does someone learn without this experience and, more importantly, without the experience of adjusting and recovering from unforeseen complications?

This process teaches even the brightest candidates to keep their head low, not take risk, and just get to the next assignment which will involve a promotion and increased compensation. If the leader exits in the midst of a long-term project, someone else will be responsible for the outcome. Normally, by the time the candidate settles into his new position and new personal surroundings, he is preparing for his next move and his next personal and professional environment. The candidate learns more about the relocation game than about becoming a leader or champion.

Defenders of this practice will argue that the candidate is acquiring valuable organizational exposure and that assignments will eventually become longer and more meaningful. This all may be true. What is overlooked is the relative ineffectiveness of the candidates at all of their preparatory assignments and the lost opportunities to improve the organization, if an experienced leader had been occupying the position. Instead of moving the organization forward, many key positions are occupied by warm bodies reluctant to make decisions, assume risks, failing to mentor or develop other employees, and simply waiting for the next assignment.

Additionally, during their young and formidable years of development the candidates spend their time with training wheels and no real responsibility. It is in the early stages of their career that people develop the characteristics they will carry throughout their professional life. It is their early environment that molds their work ethic, habits, self-confidence, and respect for peers and subordinates.

The forgotten aspect of this development process is the message sent to the larger workforce and those directly involved with the candidate's assignments as either peers or subordinates. The message is that senior management is not concerned enough about their facility or department to give them an experienced manager with a long-term commitment. The message is to keep an eye on the rookies and "we hope you can clean up their mess". A simple example is the decision by a division of an automotive company to re-hire terminated employees during a union strike and contract negotiation. This created a situation where management and HR ceased to enforce policy when it became clear that terminated employees would be returned at the end of future contracts. This was even unpopular with other union employees, who were sincerely attempting to contribute to company success. This decision was made by a fast-tracker ready to move to his next assignment. It would not have been made by a seasoned veteran.

Are there good leaders that come through these programs—sure. When you have excellent candidates entering the program there will be many who are successful, regardless of the merits of the program. Organizations with longer-term assignments, who provide experienced mentoring, and minimize training positions outside the candidate's knowledge base, are more likely to be successful.

New Procter & Gamble CEO David Taylor stated at an investor conference in 2016: "In the past, the company has groomed rising talent with a succession of rotating assignments in different business units, which resulted in parts of the operation being run by less experienced managers." He pledged to slow that down to increase institutional memory and talent.

Japan and Germany possess a better reputation for developing technical and manufacturing leadership. Their core development does not include the widespread fast-track programs based on short-term assignments.

INTERNAL LEADERSHIP DEVELOPMENT

The second type of champion development is normally practiced at smaller organizations. A long-term employee or family member who has demonstrated competency is promoted or is chosen by default when the incumbent rides into the sunset. I do not believe there is a more difficult position than that of a second-generation owner or a plant manager who has worked virtually his/her entire career at the same facility; I would consider this assignment the toughest job in manufacturing. Machining is a business that is driven by technology and is always changing. Machining is also a business driven by innovation. Successful machining companies are able to create unique fixtures, tooling, and programming, but also create human processes that deliver productivity, quality, and customer satisfaction. The second-generation owner has little exposure to alternative methods, processes, equipment and tooling. He/she has even less exposure to support functions such as installing ERP, inventory, accounting, sales, purchasing, negotiating, and ISO. He/she is likely to only have been mentored by one person. Yet he/she is responsible for a group of employees and their families and may be responsible to other shareholders and to other family members for profits.

It is just not feasible for someone with only internal knowledge to advance your organization as quickly or effectively as someone who has been with multiple organizations in a leadership position. Candidates with several assignments over five years in duration have an overwhelming advantage above the loyal but limited-exposure internal candidate.

EXTERNAL LEADERSHIP SEARCH

The third approach is to search and attract the best candidate to lead your company, facility, or engineering department from outside your organization. I would promote filling open positions at all levels with roughly a 50/50 split between internal promotions and outside hires. This provides opportunities and motivation for advancement for existing employees, while injecting new ideas and new blood from the outside world. When it comes to key leadership positions you must insure, for the benefit of the organization and all employees, that you have the top talent you can secure to move your organization forward.

I have been fortunate to manage two acquisitions where my company was able to purchase a direct competitor. This provided an opportunity to study in great detail the strategy, management, and the alternative processes to machine and manufacture similar and sometimes exactly the same product and part numbers. In this analysis, all three companies possessed primarily identical equipment. Even with identical equipment there were dramatic differences in the processes to manufacture the same item. These differences went well beyond specific machining choices and were strategic in nature. The organizational structures, quality systems, facility layouts, software, and cultures were also extremely different. Each time I was able to identify the best macro- and micro-practices to create tremendous quality and productivity synergy. These best practices were then shared with all sister facilities. This was only feasible because of the opportunity to lead multiple facilities and absorb the entire workings created by another talented team over several decades. How can this not make you a more effective leader with a much broader scope of experience and understanding?

Even at organizations manufacturing dissimilar products there is considerable expertise developed over decades that is applicable to your organization. This expertise may be broad, i.e., application of ERP software, effective approaches to ISO compliance, capacity management, lean experts/Six Sigma, or knowledge of customers and suppliers. The expertise may be narrow, i.e., innovative workholding, automation equipment, inspection devices, or material-specific cutting tool productivity. The individual without the external experience is operating with a severe handicap.

LEADERSHIP SUMMARY

The most pertinent manufacturing-leadership quote I have seen comes from a former 3M CEO, William McKnight (Brown et al., 225):

As our business grows, it becomes increasingly necessary to delegate responsibility and to encourage men and women to exercise their initiative. This requires considerable tolerance. Those men and women, to whom we delegate authority and responsibility, if they are good people, are going to want to do their jobs in their own way.

Mistakes will be made. But if a person is essentially right, the mistakes he or she makes are not as serious in the long run as the mistakes management will make if it undertakes to tell those in authority exactly how they must do their jobs.

Management that is destructively critical when mistakes are made kills initiative. It's essential that we have many people with initiative if we are to continue to grow.

McKnight discusses delegation in the context of growth, but his advice is relevant with or without growth. In an era without lifetime employment, a more migratory workforce means that organizations need to be in a constant state of leadership and technical development. This realization is augmented by the preponderance of baby boomers in key manufacturing positions, who will be exiting in waves.

Not only do our manufacturing leaders need to excel at people development, but there must be an acceptance of the digital growth, data explosion, and connectivity surge permeating the manufacturing sector. Beyond acceptance, the leader needs to be adroit at when and how his organization implements these new technologies. Timing can be as critical as the "what" and the "how". Critical projects with new customers should not fight for resources nor become victim to scope creep and implementation quagmires. Manufacturers can be digitally savvy and prepared for Industry 4.0 without wrecking operations. Implementation and new technologies absorbed in stages and through new equipment or new software installations minimize stress and risk.

Just as in football every successful team needs a quarterback who can score at the end of the game. The team may not win every game, but they know the quarterback is not going to lose the game. The team knows it has a leader who will make the solid decisions and that it will win or will have a chance to win at the end. This provides confidence. Fortunately, in the machining business there is not just one winner and one loser but many companies who can have strong profits and a winning season.

The leader needs to transcend his own career and goals. He/she needs to meet with individuals and groups to communicate a vision that enables people to be part of something larger than themselves. The common thread that ties all employees together is long-term success. My belief is that there will always be economic cycles, but if the company and facility remain strong for the next decade or the next two decades, the entire team has been successful. The good jobs will be available for the friends, neighbors, family, and community. Along the way, shareholders have been prosperous and employees have developed their careers and funded their retirement. It does not matter if there is an ownership change, because the objective is to possess a profitable facility with skilled employees, sought-after equipment and processes, and happy customers. Being purchased by a new owner is a badge of honor. In this context everyone understands that if employees and management work together, we all win and our path and goals are intertwined.

Finally, Steve Jobs was quoted as saying he would rather have a home run than two doubles. It is hard to argue with Jobs's ability to create and market dynamic new products, but in manufacturing I want a leader who not only prefers two doubles but prefers three singles. Three singles score a run just like a home run, but it requires teamwork. With the singles and doubles you still have runners on base that will continue to score with another hit or walk and you do not have to pay the home run hitter the big dollars or lose them in free agency. There are only so many home runs in manufacturing, so we need to win in other ways!

5 EQUIPMENT

Even the most successful organizations have limited capital to invest. Machining organizations do not have the luxury of striking out when installing capital equipment. There is a lot more at stake than just dollars. These projects require substantial technical resources, physical changes to the shop floor, utilities, and likely promises to customers. Machining organizations not only install new equipment for increasing capacity and cost reduction but also to simply replace equipment at the end of its useful life. As complex CNC machines age, the maintenance costs move in one direction, while the productivity moves in the opposite direction (see Figure 5.1). Equipment selection and installation simply need to become an organizational core competency. You better have a leader and manufacturing champion who is capable of heading a team for determining what to buy, how to buy, and how to install. Productivity and precision are not linear developments in the CNC machining environment. Frequently, a leap in productivity or accuracy is generated not from the machine itself but from one or a combination of the corollary machining processes as shown in Figure 5.2. An acquisition team that understands the

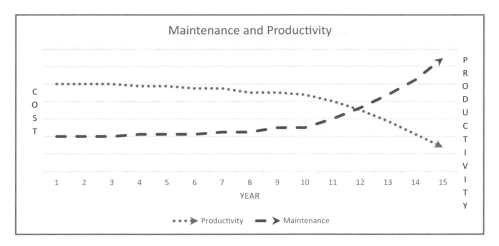

Figure 5.1

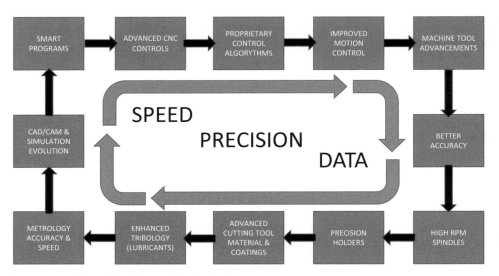

Figure 5.2 Synergies of Machining Technology

holistic technical process upstream and downstream of the chip separating from its parent superstructure is needed to maximize the positive impact of a capital investment. In essence, the investment can be leveraged or magnified with the synergistic integration of the scientific and engineering developments of the adjacent technologies.

The most important aspect of your equipment is that it is matched to your products. By "matched" I mean optimal as it relates to process, quality, productivity, automation, ergonomics, and reliability. Of course, your equipment needs to be large enough, fast enough, and accurate enough. But the equipment frequently is also too large and too fast. We have all seen 90" machine tables with an 18" part being machined. Overall, this is very inefficient and a poor use of capital. The large machine has a large maintenance budget, a large footprint, and a large price tag. The large spindle has low RPMs, and the large table has slow rapid rates.

You should seek large work for your large equipment. Rather than run significantly undersized work on larger equipment, you should purchase a smaller machine or preferably add a nightshift or additional nightshift personnel. Machining smaller work on large equipment is much more difficult for the machinist. There will likely be much more leaning or climbing. The result will be poor ergonomics and safety, and slower production.

Employing over-sized equipment is frequently less obvious than in the case of the afore-mentioned 18" part on a 90" table. A common example is a six-inch component centered on a 24" horizontal machining center, which will require approximately nine inches of additional tool reach, if you need to rotate ninety degrees to machine either side of the component. This additional reach will generate deflection, poor tool life, increased tool runout as well as slower cutting speeds, and perhaps it will be the determining factor for high scrap rates or the inability to hold tolerance.

Likewise, a small part on a medium-sized five-axis vertical machining center can experience critical clearance issues for the tools to reach past the turntable or the workholding. Again, this may be the determining factor for quality, scrap, and productivity.

So how can a machine be too fast? It is not necessarily that the machine is too fast but a question of what the machine designer sacrificed to achieve the speed. Normally, fast machines refer to high rapid rates and high RPMs. The adjective "high" is subjective and always changing. It depends on the type of equipment. Without angering the OEMs it is sufficient to state that there is always some degree of tradeoff between rigidity and speed. You might say the laws of physics are at work. If your application is a combination of material hardness and geometry that dictates unavoidable cutting forces, then you need a machine tool suited for rigidity. If not, your machine life will be reduced, your maintenance costs will jump, downtime will leap, and the high RPM spindle will lose its accuracy. Replacing a few high RPM spindles will blow most maintenance budgets. There are programming methods that utilize lighter cuts at higher feed rates that assist with these issues, but this method cannot completely compensate for a poor match of machine to application.

Conversely, if your application requires small cutting tools to machine aluminum components, you need 20,000 plus RPMs to be successful. Rigidity and torque will not be issues.

There are many more aspects of matching your equipment to your products than just size, speed, and accuracy. The first step on your decision tree when purchasing new or used equipment is how can you innovate to implement a type of machine or process that is revolutionary rather than evolutionary for the application. If you can turn a product family to size and surface finish on a lathe, while your competition must grind, then you have hit a home run. Do you need a mill/turn or a turn/mill machine? Can you bore instead of jig grind or ball nose mill versus RAM EDM? Do you stack parts on a wire EDM or machine individually? Should you hard mill or mill to size soft with heat treat compensation factored? The bottom line is that if you have the correct equipment, you can compete, and if you do not have the correct equipment, you just better hope your customer does not find the company that does possess the correct equipment for the application. If you have the correct equipment, you will need to build and execute the rest of the Machining Pyramid, but you will be in the game. If the competition has equipment that is much better matched or more innovative to the product, you will probably not be in the game.

For example, I needed to produce two stainless steel components for a medical instrument for which the competition was utilizing three lathe operations and two mill operations per component. Total volume was approximately 600 pieces per month for each component. Both components were very similar with only minor differences in geometry. After securing the business, I established a cell of three lathes and two vertical machining centers. Two machinists operated this cell, with one running the lathes and the other the machining centers. We developed very innovative gauging and performed all the inspections within the cell. Despite multiple improvements in the cell we experienced low margins, about 4 % scrap, and constantly fought deliveries. We essentially were utilizing the same equipment as our competition. Since our price needed to be lower to obtain the business, there was just no profit. We held a Kaizen event to determine our next step. We purchased a new lathe with live tooling, twin spindles, twin turrets and a bar feeder. This meant that we could do one part completely on one machine. We had a single piece flow with virtually no scrap. Margins were terrific and deliveries were ahead of schedule. We re-deployed the three lathes and two vertical machining centers to other work. We were even able to set up a programming station for our machinist to generate lathe programs for other equipment. Any organization without a similar piece of equipment would never be competitive with the new benchmarks we established for these components.

MACHINE ACCESSORIES

The accessories that you have equipped onto the machine tool are as important as the type of machine tool itself. There are frequently dozens and sometimes hundreds of machine and control options. These choices may significantly improve the quality and productivity of your operations. There will be times that your competition has the same piece of equipment, but the choice of options will be the difference between winning and losing. There are the obvious selections of spindle RPMs, glass scales, coolants (or oils) through the spindle for mills and through the tool for lathes, etc. You will need to determine whether to install a tool setter, spindle probe, and/or thermal compensation. There are control options ranging from the amount of memory to tool-life management to work setting error compensation (WSEC).

Each type of equipment will have its own set of options to evaluate. A grinder is very different from an EDM, which is very different from a lathe. Likewise, there are many different types of grinders, many different types of EDMs, and many types of lathes. Each type or category of equipment will have subcategories. You must be able to evaluate and equip each purchase successfully to optimize your equipment. If you believe you have learned from previous mistakes and will do better the next time, it may not be true. If you think you were successful accessorizing your last purchase and will be successful again, it may not be true. It may not be true, because technology is alive and well in the machine tool industry. Machine designs, controls, software, and third-party options are constantly improving. The combination of machine type, control, and options you specified previously are likely no longer available. If the combination is available, it is probably no longer the optimal combination due to better offerings, or your needs have grown.

This brings us to the point of standardization. Should you order your last combination of machine type and options again? The answer is not a simple "yes" or a simple "no". Standardization is good, but standardization without updates is bad. Standardization for the sake of standardization is worse. If you are equipping multiple facilities at the same time for the same product, then standardization is good. If you are buying several pieces of equipment at the same time for the same facility and for a similar purpose, then standardization is likely to be good. If on the other hand you have been buying the same brand of machine with the same control and same options for years, then your standardization may be stale. It sounds good to have common controls, common spare parts, and common service. However, this fairytale normally does not last very long. The same machine builder with the same control does not stay at the pinnacle of their industry more than a few years. You will invariably find yourself sacrificing a more sophisticated machine with new options and features for the sake of your standardization. Shopping the competition is likely to yield a better product for a better dollar. At some point this gap will be too great to justify your standardization. I will bet that your distributor and your manufacturer are so sure of your business that you are not even getting the best price any longer.

The acquisition process should also include an understanding of how you are going to create your CNC programs. If you are acquiring equipment similar to existing machines and controls, this may be a simple step. If you are acquiring a type of machine, a type of control, or additional axes or features not present in your shop, then this step may be complex. The normal options are to pay for turn-key programming, to update your CAM and post-processing, or to program on the shop floor. This decision needs to be made in the acquisition phase, as there are frequently control options and software to facilitate shop floor programming. Whatever your decision, it needs to be included in your budgeting and negotiations. You need your training and software in place, so that the machine is not idle after installation. Most importantly, everyone needs to understand that the new piece of equipment will not be optimized and will not achieve a quick ROI unless the programming capabilities are sufficient. Some higher-volume companies will only need to program one or two parts for their new machine. Other companies will be programming dozens and eventually hundreds of parts. Either way, it is not unusual for programming and the debugging of those programs to be the bottleneck with new equipment.

Similarly, you need an understanding of how you will inspect components machined on your new piece of equipment. Many types of machines now incorporate methods to inspect parts on the machine or at least certain features on the machine. Again, this needs to be included in your budgeting and negotiations. It is generally twice the cost to add a feature onto a machine after delivery than negotiating that feature at the time of purchase. Whatever features of your components that are not inspected on the machine will need traditional inspection methods. The size of your parts accompanied by the quality and size of your inspection equipment must be incorporated into your layout. As we discussed with programming, you will struggle with achieving a ROI, if your machine is idle while your team is trying to determine whether the parts are within print tolerance.

AUTOMATION

We have talked about the type of equipment and the accessories on your equipment, but we have not talked about automating your equipment. There is an old saying: Emigrate, automate, or evaporate. I do not know who to credit for this quote. Since I do not want you to emigrate or evaporate, I want you to automate.

It is a given that the higher your volume the easier it will be to automate. Higher-volume organizations may be able to purchase equipment that is specifically designed for their component(s), enabling a high degree of utilization, efficiency, quality, and automation. The fact that your shop is not high-volume does not mean you should not try to automate. Actually, it is more reason to automate. If you are successful, you will likely have an operational excellence advantage over your competition.

Automation by itself is not difficult. Everyone understands that a robot or other device will function as intended. Within the machining environment it is common to automate high-volume, open-tolerance, and soft material applications. When the application is close-tolerance, hard material, and/or low-volume, then success is much more difficult. The automation is not the challenge. The challenge is creating a non-human closed-loop process capable of controlling the rapid tool wear due to hard material and maintaining the features within the print tolerance.

I was able to create a machining cell for medical instruments that accomplished this challenge. The cell utilized robots to load small pallets to five-axis machining centers. Success was only achieved by a talented and diligent technical team that validated and refined the programs, established laser measurement and offsetting of all tools, employed spare backup tools, probed pallets for accuracy, designed innovative workholding, simulated programs, utilized stubby and balanced holders, WSEC, and CMMs adjacent to the machine tools. There were many additional subtleties necessary to achieve unattended operation. This success was replicated across many cells for additional part families.

Automation will also force you to standardize many more aspects of your operation than you would anticipate. This standardization will require communication, teamwork, and problem solving. All of these are traits of world class companies and all are goals which to aspire. A byproduct of automation will be that your top machinists may be better utilized. Instead of portraying that the automation is replacing operators or machinists, I believe it is more accurate and certainly more palatable to articulate that you are leveraging the knowledge of your talented workforce to set up unattended machining superior to your competition. You are giving your best people an opportunity to shine, to beat the competition, and to ultimately maintain high wages and benefits through productivity and innovation. There is no other way for long-term success against developing economies with wages and benefits 1/10th the Western economy.

Without automation the best machinists are assigned to install and debug the new equipment, but then must spend the next ten years operating the equipment each day. With automation, the best machinists can be assigned to install and debug new equipment and then move to the next challenge, while employees of lesser skill attend to the daily requirements.

Automation projects can be challenging for high-volume companies with deep pockets and a large support staff. Automation projects can be frightening and risky for small to mid-size companies with low-volume applications and smaller staffs. Automation is more complex when the hardware and software needs to accommodate a higher diversity of part numbers.

Let us take a few minutes to define automation and put some readers at ease. Robots and transfer lines are certainly a category of automation. I would consider these to be a home run for their level of reducing human involvement. Better stated, they are home runs for increasing productivity per hour worked. There are many ways to incorporate automation successfully without hitting home runs. A few singles and a double can win the game. A bar feeder on a lathe is very affordable technology and can automate the load process. Accompanied by a parts catcher or parts conveyor, the lathe can run unattended for hours. Tombstones can be set up to hold dozens of components that will yield hours of unattended machining on a Horizontal Machining Center (HMC). Most HMCs will accommodate two tombstones. Each tombstone has four sides, and you can configure parts on top of the tombstone to create ten locations. You can load the same part on all locations or different parts on each side of the tombstone. Whatever your configuration, you have the opportunity to create long unattended machining cycles allowing your machinist to complete other activities. With some ingenuity, most equipment types can have multiple parts that are positioned within the working envelope concurrently to significantly expand the unattended machining time. Most of these machines can also be configured with tool-break detection, tool setting, tool-life management, or basic counting of machined parts with programmed tool offsetting.

...y mechanical, electronic, or programmable device that allows your equipment to operate without the presence of your machinist is providing you with some degree of automation. This will allow your machinists to run other equipment, inspect their parts, or prepare external setup activities for the next job. In other words, you are utilizing the machinist's talent more effectively and this is the only way to defeat low-cost countries.

Every decision regarding your equipment must be designed to allow your high-skilled people to either operate multiple machines, perform value-added activities, or to perform unavoidable work while your equipment is making chips. I am using some of this terminology for our Lean readers. We certainly do not want to automate so that our employees can sit on their behinds and watch. However, you need to understand that it is better for the machine or machines to run while the employee sits on his behind than for the employee to work off his behind and the machine not to be making chips. The hamster running faster in the wheel is not a good strategy for a modern machine shop. Optimally, the machine or machines are running, and our people are inspecting, programming, changing tools, or performing other necessary activities to keep the equipment producing good parts or preparing for the next setup.

How does a company or an operations manager know when to assign a machinist to operate one machine tool and when to assign the machinist to operate two machine tools? If the wrong decision is made, either the machinists will not be productive, or the machine tools will be sitting idle, because the machinists are too busy loading, unloading, inspecting, documenting, and problem solving, or the machinists will produce scrap, because they were too overwhelmed with manual activities to control two processes concurrently. Since most machining organizations are not high-volume, the decision variables will change every time the machine is set up with a different part number. The correct decision today may not be correct tomorrow.

Through research and empirical analysis of data I have determined the process variables that dictate the amount of time required by the machinist or the operator to properly attend a given machining operation. Creation of a formula and a simple program allows the process variables to be entered and an attending ratio to be calculated that provides a quantitative measurement of the human time required to support the single machining operation. When considering assigning the machinist or operator to run a second machine, both machining operation variables may be entered to determine a quantitative measurement of an attending ratio to run both machines. The attending ratio is charted to predict the utilization of each machine tool based upon the combined process variables. Operations management can decide, if the projected machine utilization is acceptable. If not, operations management may choose to alter the schedule to select machining operations that are compatible, providing that the customer deliveries are maintained. The attending ratio is applicable to all machining operations and can be readily adapted and customized to any location.

Ultimately, if you have made good decisions regarding your equipment and processed your jobs properly, the equipment can make chips, if the machinist is gone for any reason. Besides completely unattended, this includes lunch, the time between shifts, or a simple trip to the restroom or inspection department. If this only generates .5 hours of extra cutting time per machinist per day, it amounts to 25 hours of chip time for a fifty-person shop. That is equivalent to three additional machines or three additional machinists. Since it is difficult to find and train qualified personnel, I would highly recommend leveraging the ones you do have and focusing on engineering your equipment and processes to keep the spindle cutting.

Automation is also beneficial for the workforce. People receive additional training, are exposed to the latest technology, and become more valuable to their current organization or future employer. When the productivity and profit of the facility is higher, the wages and total compensation can be increased. Replacing and retraining people becomes longer and costlier, and hence employers will seek to reduce turnover. The modern workforce comprehends the necessity to automate in order to be able to compete globally. If there is a lack of understanding or motivation within the general workforce for automation, then leadership needs to communicate and educate more thoroughly.

ACQUISITION TEAM

We have discussed buying the right type of equipment that is matched to your product and outfitted with the optimal accessories and automation. The question before us now is "how" to analyze, test, negotiate, install and debug this equipment without breaking the bank and without breaking your employees, who already have a fulltime job. The first step is to realize that this process should be fun! Yes, I did say "fun". Your leader should assemble a team and let it help define the goal of the project. In the Lean vernacular this could be a series of Kaizen events. The leader should set a tone and spirit of teamwork, competitiveness, esprit de corps, and innovation. It is time for the team members to use their talent and creativity to revolutionize a process or product and put your company ahead of competitors for years. Every machine acquisition has an opportunity to leapfrog the competition. Why look for 10% improvement when you can reach 50% or 60%? Aim high! The individuals and the team do not have the opportunity for this type of event often, so it needs to be enjoyed. Using your experience and creativity to beat the company across town and across the globe is fun.

I want to emphasize that it does not take a miracle to reach sixty percent improvement. Adding options or workholding to create more cycle time in conjunction with altering your layout to facilitate operating multiple machines is a huge advancement. The Kaizen event coupled with the purchase of new equipment should be the catalyst for physical and cultural change. This is just one of many methods to leap forward, and that is the beauty and the beast of machining and manufacturing.

Are you doubtful that you can create enough unattended cycle time for your machinist to run two machines? I am not talking about manual machines such as a Bridgeport mill or a manual surface grinder. I am not referring to a large boring mill located one hundred feet away from any other machine. As already discussed, there are many options to automate the worn tool replacements. Spindle probes will locate and inspect parts. Whether your team is purchasing a grinder, lathe, or EDM, there are multiple approaches to creating unattended machining time, and that needs to be their clear goal. As you can see from Figure 5.3, in order to compete against low-cost regions it is normally a necessity to divide the machinist's labor and the overhead across multiple machines.

Generally, the team should consist of a CNC programmer, a manufacturing engineer, one or more machinists, and an operations manager. The team should contain some of the people responsible for successfully installing, debugging, and operating the equipment. The leader of the team should be your technical champion. The technical champion needs to be adroit at facilitating meetings, brainstorming, and managing projects. Most importantly, he needs to instill a vision into the team and keep them aspiring to hit a double, triple, or home run. Capital projects should include a definition and list of machine and project specifications. The more custom the project, the more detailed the list. This helps to achieve consistency between quotes and analysis of the quotes. Most teams will create some type of matrix listing the various manufacturers, price, important features, etc. Regardless of your organizational commitment level to Lean and Six Sigma, the technical champion needs to apply lean concepts/Six Sigma concepts to the project. If you decide to have sample parts machined, I strongly recommend having someone present to view the setup and machining. It is also preferable to view the programming and to bring the raw material in person to the runoff, instead of shipping ahead. This permits you to view the total time and total complications associated with producing a good part from scratch. If you cannot be there in person, an option is to laser your logo on the raw material and ask for all your parts to be returned. It is important to understand, if one part was machined to yield one good part, or if five parts were machined to yield one good part.

You should have your best people involved at this stage of the process. It is not uncommon for an excellent application engineer to put lipstick on the pig and an unprepared application engineer to hide the beauty queen. In other words, do you know which machine tool performed better or just which application engineer was better?

REGIONAL COST COMPARISON

MANUFACTURING SOURCE	SALES PRICE	DIRECT LABOR	MATERIAL	OVERHEAD LABOR	TRANSPORTATION	TAX & REGULATION PENALTY	QUALITY PENALTY	TOTAL COST	PROFIT
WEST	1000	250	150	350	0	75	0	825	175
WEST - SPLIT LABOR*	1000	125	150	175	0	75	0	525	475
LOW COST REGION	1000	50	125	200	100	0	200	675	325

*COMPETITIVE WHEN MACHINIST RUN MULTIPLE MACHINES

Figure 5.3

There are many logistical and contractual issues unique to every project that I will not discuss. There is one important approach that is common to every project that needs to be emphasized. Once you have chosen a manufacturer, you should have a long meeting with their best technical personnel and your team to review every standard feature and every special option available for both the control and the machine. I refer to this meeting as the final order review. The manufacturer is very cooperative when they understand that they will receive the purchase order shortly after the meeting. This meeting will help to insure that you are not missing a critical option and that you are aware of all features that may help your quality and productivity. The reason you need to perfectly comprehend all standard features is to insure that you do not buy an unneeded option or conversely fail to buy a needed option. If you save dollars by avoiding an option, those dollars can be redeployed elsewhere in the project. Additional capital for inspection equipment, workholding, tool holders, etc. can be critical to the success of the project. It is beneficial to include one or more of your machinists who will be operating the equipment. Their input is important, and instilling a sense of ownership is an important factor to the overall success of the project, as this additional commitment level will speed the learning curve or overcome small or large future technical hurdles.

There has not been one instance where I or my team did not make an important change during the final order review. By incorporating additional technical resources from the manufacturer you will receive better recommendations and a more thorough technical review. The final order review also serves to lock in your entire team to the final choice of options and features. In other words, you are all jumping off the cliff together. This can be very important during those dark days on a difficult project. The entire team will be more focused and determined to fight through severe challenges, because everyone was involved in the decision-making process.

INSTALLATION

High-volume applications simplify installation and startup, as the project will either be turnkey or involve a small number of similar components. Conversely, low-volume manufacturers may have hundreds of components to program and debug over a long time period. The complexity of the low-volume installation is enhanced, because many machine or software options may not be required for weeks or months until the components requiring the specific option are on the machine. Similarly, the variety of components for the low-volume manufacturer may not expose an equipment deficiency for some period of time. Different geometries will require a combination of rigidity and accuracies over a different range of RPMs, sizes, and axes.

Below are some general best practices for installation:

- Layout: Involvement from the entire acquisition team with key input from maintenance and the manufacturer to insure your floor, mounting, utilities, and environment will not impact warranty. Many manufacturers require excessive physical installation specifications for their equipment. If you are not going to comply, I suggest you have a written trail to insure your deviation, no matter how practical, will not be cited as a reason for the manufacturer to avoid warranty claims. The layout must also be conducive for material flow both upstream and downstream as well as facilitating machinists operating multiple machine tools. Layout complications are a significant reason for project delays and ROI deficiencies.

- Runoff/Artifact: High-volume applications require X number of pieces normally both at the OEM's location and after installation at the buyer's facility. The runoff normally specifies a CPk and time per piece. Lower-volume applications require more creativity. It is advantageous to create an artifact that incorporates the more difficult geometries and tolerances representative of the overall population. Successful machining of the artifact is indicative of future success. Additionally, more artifacts can be machined at any point in the future to compare the machine to its original condition or to compare machine-to-machine accuracies for both existing or future equipment. It is necessary to create a standard raw material condition, workholding, tooling package, and inspection method to isolate any variance in the artifact to the machine tool.

- Terms: The most prevalent industry terms are 10/80/10. This may vary or be altered for a number of reasons. I have found the most important aspect or terminology to be associated with the final payment. Utilizing the terminology, 10% down, 80% upon receipt, and 10% upon successful installation is beneficial for the buyer. There will be many unforeseen circumstances, and incorporating "successful installation" allows the buyer to define success during these unforeseen conditions. The OEM needs to be a long-term partner, hence reasonable interpretation and integrity is important. The buyer will need service for many years, so maintaining a positive relationship with the distributor and OEM will maximize overall ROI. Still, maintaining leverage over your capital seems to bring projects to fruition in a more seamless manner.

- Signoff: There is a wide range of installation verification procedures practiced by OEMs. If it is not offered, then the buyer should request a written analysis of the accuracy and runout of each axis. This should be performed after installation on the buyer's floor. Additionally, the service personnel should review each option and verify that the option is functioning as designed. It is very common for optional features, whether standard options or special options, to not have been installed or to be not functioning in harmony with other features. This verification may prevent long and difficult problem solving.

CONCLUSION

Is technology changing faster in the machining environment than in other industries? I believe the answer is "yes"! Advancements in the material sciences consistently yield improved cutting tools both in the base material and the coatings. Advancements in chip technology and processing speeds yield more sophisticated controls. Advancements in software enhance CAM systems which produce more precise CNC programs that are interpreted by more sophisticated control software and control architecture. Advanced materials, sensors, and motion control lead to improved machine tools. Advanced fluids (tribology) provide improved coolants and lubricants. When the cutting tool, wheel, or electrode hits the metal there are tremendous synergies between these sciences. Frequently, an advancement in one field unlocks the latent potential in another. High RPM spindles were far ahead of the cutting tool capabilities until improved carbon compounds with six layers of coating were introduced. Advancements in multiple fields propel the machining industry by leaps and bounds. Those who combine these factors in their machining applications will outpace the competition.

The machine tool industry is fragmented among hundreds of OEMs dispersed globally. New models are introduced annually. Our machines do not carry passengers on the ground or in the air. Hence, new concepts and technologies are implemented quickly. I do not believe there exists a more dynamic industry encompassing so many fields of science. The ability to select and implement the machine tools for your current needs and future vision needs to be a core competency. Most organizations simply cannot afford to swing and miss on decisions of this importance.

6 INFORMATION TECHNOLOGY

Machining has evolved from manual machine tools with paper blueprints to a cradle-to-grave digital format. The digital birth occurs when the designer finalizes his solid model. This model becomes the centerpiece of the manufacturing process. Each progression of the manufacturing process utilizes this three-dimensional solid model to improve the speed to market, quality, and productivity of both the process development and also the manufacturing process. Since this model represents the perfect digital component or the perfect assembly, each step of the manufacturing process strives to exactly reproduce the model in the physical world. In this context, the manufacturing processes revolve around the solid model and are dependent on the solid model similar to the planets' reliance on the sun.

Each manufacturing process in turn has digital outputs linked to them similar to the moon(s) around a planet (see Figure 6.1). Each solid model generates dozens of digital outputs, and each time the manufacturing process physically brings this model to life through a production order, an endless stream of process data is generated. In fact, every CNC machining operation will require 10–15 electronic files consisting of CAM files, programs, setup sheets, metrology programs/instructions, and data. Within this process data lie the seeds of knowledge to improve quality, productivity and profit. In this analogy every 3D solid model is a sun in its own galaxy, with the dependent digital planets relying on the sun for their existence. Collectively, these galaxies create massive process data that must be stored. It is your choice whether this data will become a black hole where nothing ever escapes, or a worm hole of rich metrics for decision making that connects all process outputs and lets you transport your organization from where you are now to where you want to be in the future.

It is hard to predict where the Industrial Internet of Things (IIoT) or Industry 4.0 will take the manufacturing community. It is evident that IT, in one form or another, is ubiquitous in the second decade of the 21st century. Evaluating and installing hardware and software will only grow in significance.

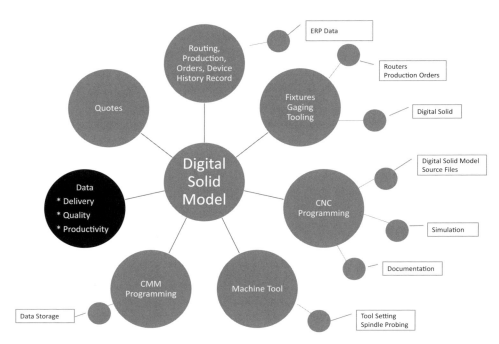

Figure 6.1 Digital Galaxy

In a modern CNC environment, software is the driving force in the IT world. The hardware is Robin, but software is Batman. The entire IT infrastructure is a necessity, but it is the software that should match your business and technical requirements. It is the software that should be mastered and customized. And it is your software that should interface with your customers, suppliers, and internal operations.

Following is a brief description of some of the common software applications required for machining or manufacturing organizations.

COMPUTER-AIDED DESIGN (CAD)

If you are an OEM, it is obvious why you need a CAD package and why it is important to your company. Most organizations own more than one brand of CAD software. This can be for several reasons. First, high-end software is not needed for many simple applications. Why spend 35K plus annual maintenance for a co-op or intern to edit existing 2D fixtures or floor layouts. Second, many companies inherit different CAD software via acquisitions. Finally, most companies transition gradually to a new CAD package and maintain some seats of their legacy software to edit existing drawings. It is generally very costly and lengthy to switch CAD software. Hence, whether you initiated the decision for a new CAD package or you inherited multiple packages through acquisitions, your IT will maintain multiple packages for years.

Use of associativity can astutely reduce both design and programming time for OEMs who control their design. Associativity systemically ingrains standard practices into both design and manufacturing, as the methodology yields consistent drawings, CNC programs, and tooling packages.

Contract manufacturers own one or more types of CAD to edit customer models and speed product design such as mirror imaging. It may be beneficial to own multiple packages to facilitate manipulation of customer models.

COMPUTER-AIDED MANUFACTURING (CAM)

Similar to CAD, most companies employ more than one CAM package. Each CAM manufacturer generally aims to support many types of machining, but most have a forte. Some CAM packages have leadership positions in five-axes milling, while others lead in multi-axes lathes. Still others excel at EDM. Hence, some machining organizations choose to have the best CAM solution for each application. Acquisitions and a need to maintain legacy software for editing old programs are the other reasons for multiple packages.

Many OEMs single-source in order to create a CAD/CAM solution with Product Data Management (PDM) and/or Product Lifecycle Management (PLM) to accomplish and the afore-mentioned associativity. The PDM application assists in controlling the ECN process and the sharing of files across multiple sites. PLM is an advancement of PDM that manages all related documents of the project.

CAM systems provide many benefits beyond the basic CNC programming function. Viewing of the model reduces the time required to quote a new job and increase the accuracy of a quote. Viewing the model during a quote helps to insure that features are not overlooked or misconstrued. CAM systems can be utilized to generate a CNC program with a process time estimate for each individual machining operation that reflects the time required to transform the raw material into a finished product. My favorite secondary use of the CAD/CAM system is to allow machinists and inspectors to view and manipulate models during the first pass through the production process. This helps to insure that they do not omit or misinterpret hidden, small, or unusual features on a complex component with multi-page prints, views, and sections. Viewing of models can be facilitated by installing viewing software in strategic locations on the shop floor and in final inspection. In addition to reducing mistakes, this methodology saves time, as supervisors, machinists, inspectors, and others will spend less time studying prints and discussing interpretations. Your people will have confidence in their analysis of the print and may find errors or improvements in the process before they will occur. Doubt costs money, whether or not it leads to scrap or rework. Models are available, so why not utilize them to your maximum benefit.

I also want to emphasize that your CNC programs are intellectual property, just as the part design, print, and model are intellectual property. A CNC program that has been debugged, refined, validated and articulated with setup instructions may be a more significant investment than the design of the component. It is thornier for competitors and wannabe competitors to reproduce your production methodology than to reproduce the design. It is very rare that a competitor cannot get their hands on any of your components and reverse engineer. However, quite often the competition cannot duplicate the cost, quality, and life of your components.

An example is the tooling that produces the opening on the ubiquitous aluminum beverage can. The aluminum lid is approximately .008" thick and must not leak prior to opening, yet must readily break free and pivot downward when depressed by the tab during the opening process. The cut or score in the beverage lid must be controlled within .0006" to properly balance no leakage, with the ability to open without first breaking the tab. The tooling that creates this score in the lid is termed a score punch. The lids are produced and scored at a rate greater than 700 strokes per minute. A single score punch can produce 30 million lids and sells for less than $1000. The blueprint is readily available, but as yet there are only a few machining companies capable and willing to produce these components. The lost production time to the lid manufacturers of a breakdown in the press is extensive. Downstream, the value of lost product to a filler who discovers leaks or opening failures can be seven figures.

ERP

Even the smallest organizations require Enterprise Resource Planning (ERP) software. We take for granted that larger organizations require a centralized system to integrate disparate business processes and personnel. I would argue that smaller companies with a lower percentage of overhead need ERP just as emphatically as their larger brethren. Companies who install systems when they are small find the task much simpler than medium-sized companies who are forced to install systems in emergencies when they discover they can no longer effectively operate their larger and more complex business. ERP was once the domain of larger organizations. Newer systems require less IT support, permit faster implementations, and are available in cloud-based solutions.

Regardless of your size, you must be proficient at ERP to operate a modern business. Customers manage cash flow carefully and request more frequent shipments of smaller quantities. Product differentiation is greater and product lifecycles are shorter. These trends have forced the supply chain to manage more transactions in a leaner environment. ERP will collect, store, and facilitate sales demands, manufacturing requirements, purchasing, inventories, shipping, and accounting functions. The data is easily retrieved and formulated for real-time decision making. Besides automating many business functions, ERP provides the business intelligence on profits, growth, and deliveries that assists leaders and managers in understanding what is working and what is not working, and in discerning winning products from losing products.

The ERP system will track estimated cost, standard cost, and actual cost. You must understand your cost when you quote projects or re-quote repeat projects. If you do not know your cost, your competition will eventually take the profitable work and you will have the losers.

As we will review later, your ERP will perform complex scheduling for all purchase orders, production orders, and manufacturing sequences.

For the reasons discussed, and many more, it is obvious that optimizing your ERP system needs to be a core competency. The organizations who leverage the knowledge within their ERP system and have developed a lean ERP environment have a distinct advantage on their competition.

TOOL CRIB

The largest expenses for machining organizations are normally direct labor, overhead, raw materials, and perishable tooling. Even small shops may consume hundreds of unique tooling SKUs during a year, and medium shops thousands of SKUs. High-volume shops will utilize a much smaller variety of tools, but in a higher quantity. This high-volume scenario makes it easy to maintain adequate tooling stock and normally provides volume leverage to obtain optimal pricing. High volume may be supplied closer to the point of use with vending machines. Low-volume/high-variety manufacturers will require much greater SKUs and will not have nearly the pricing leverage.

The critical issue of managing your perishable tooling inventory is to have the necessary tooling when it is needed. If you have a stock-out situation, you may not be able to run specific orders and equipment for multiple shifts. Productivity and delivery can be seriously affected. Frequently, management will force a change to the process to utilize available tooling that is in stock in order to prevent excessive downtime or late shipments. This consumes your engineering and programming resources while inducing variation into the process. There is a higher chance of scrap/rework due to the introduction of new CNC code and new tooling.

Many machining organizations have little or no visibility to customer demand. Some components never repeat and some do not repeat for years. When orders do repeat, there is no guarantee regarding the quantity and hence no prediction of how much perishable tooling is required to complete the order. One common factor of delivery, quality, and cost is availability of tooling. Common tooling may be available in days, but less common or special tooling may take weeks or longer. The solution is tool crib software that provides a solution specific database, purchase orders, inventory control, and reports.

Selection, standardization, and management of the perishable tooling process significantly affect CNC programming, purchasing, and manufacturing. They also directly affect productivity and cost. This is one of the most overlooked necessities by machining leadership.

QUALITY RELATED SOFTWARE

Many quality functions require significant use of complex software packages. Coordinate Measuring Machines (CMMs) require programming software to interpret 2D or 3D electronic files. A programmer with intricate metrology knowledge manipulates or creates the electronic file to select which features to measure, creates the datum structure, tolerances, probe selection and reporting requirements. The eventual result is an error-free automated program for the specific part number and specific stage of production corresponding to the manufacturing router.

Statistical Process Control (SPC) requires separate software that accepts various forms of electronic data, manipulates, graphs, and stores the raw and manipulated data. Most packages provide a complete suite of statistical and graphic tools. We will discuss SPC in more detail in the in-process inspection section of our quality foundation.

All manufacturing facilities contain hundreds, if not thousands of pieces of measurement and laboratory equipment that must be documented and calibrated. This includes employee-owned devices. This task is only feasible via calibration software that tracks all devices and records ISO required data specific to each instrument. This not only includes the how, when, and what to calibrate but the results of the calibrations. If a device is out of calibration, the magnitude and direction should be recorded. In some cases, customers will require an analysis to determine whether shipped components approved on the basis of the instrument found to be out of calibration were actually within specification. This can only be accomplished by knowing the details of the measurement error for the instrument. Typically, all instruments are labeled with some type of gage number and the date for the next calibration. The quality department will insure that each instrument is obtained, calibrated, updated within the database, and will apply a new label.

CNC PROGRAMMING AUXILIARY SOFTWARE

The use of CNC simulation software such as Vericut continues to expand. These packages are designed to reduce setup and debug time on the shop floor. They will also assist in preventing crashes and will compare your digitally machined component to the solid model. Preventing a crash, wreck, bump, or whatever your terminology may be for an unplanned collision, carries large ramifications. High RPM spindles cost between $15,000 and $40,000 to replace. There may be more severe damage to your machine tool and long-term accuracy concerns, depending on the severity of the mishap. There will also be significant costs due to the downtime of the equipment. The combination of reduced setup time and accident prevention achieved through simulation is huge. All the software and modeling trends are moving in the direction of improving and mainstreaming the simulation tools. If you are not utilizing this approach on your more complex setups, you should investigate. This technology is particularly beneficial on four- or five-axis machining with tooling reach and clearance issues.

Your CAM system requires machine-specific post processors to output the CNC code in a format that each machine tool will properly execute regardless of control type, age, control options, or accessories. Post-processors may be purchased from the CAM provider, a third party, or some senior programmers can develop them in-house.

OTHER SOFTWARE

There are dozens of additional software requirements within the normal operation of a manufacturing organization—regardless of size. These range from accounting requirements for payroll over sales/marketing software to track sales calls, quotes, expenses, etc. There are software requirements for overnight shipping, preventive maintenance, spindle probe programming, and a host of business and administrative functions.

DOCUMENT STORAGE

Many types of industries are required to maintain massive amounts of documents. Machining is similar and in some ways perhaps more challenging. Top organizations utilize all employees to generate documents as part of our lean doctrine to push decisions to the lowest level. Most machining organizations are operating multiple shifts and weekends. A multitude of documents are created around the clock. A listing of some of these is shown below:

Table 1 Documents

Calibration	Inspection results
CNC program	Preventative maintenance
Coating certifications	Purchase orders
Device history records	Quality work instructions
Digital pictures	Quotes
Drawings	Raw material certifications
Final inspection forms	Routers
Heat treat certifications	Setup sheets
In-process inspection forms	Tool crib
In-process inspection instructions	Tool sheets

The complexity in machining stems from the fact that these documents need to be stored (write-access) by all employees, but should be "read only" by all employees. In other words, there should not be edit or accidental delete or move capabilities. These documents need to have a file name that discerns part number, revision level, operation number, and perhaps other information. The average organization will have thousands of similar documents in each category above that will never be retrievable without exceptional discipline in directory structure and file naming. This is normally accomplished by a temporary directory, which is monitored by just a few people who then confirm the file name and transfer to a read-only directory.

IT SUMMARY

The Machining Pyramid promotes the seamless electronic integration of all business functions. Consider the cradle-to-grave electronic pathway a machined component travels, regardless of type or size of machining organization, as discussed in our galaxy analogy. We start with a digital model and proceed with a quote (make vs. buy), routing, BOM, fixture and gage design, CNC program, ERP scheduling, machining, CMM inspection, shipping, billing, and metrics.

All of these are electronic functions that require software, human training and interface with the software, and document management. The flourishing machining organizations need to integrate these handoffs as a relay race on a track team passes a baton. Those companies that master the handoffs and leverage the digitization and real-time availability of information will be the winners. The Industrial Internet of Things (IIoT) and Industry 4.0 will continue to advance these concepts. In the meantime, you may have the dreaded islands of automation. Again, do not let perfection get in your way of progress. It is better to have an island of automation than no automation. You will be able to connect these islands going forward.

Faster and smarter CNC controls, improved CAD/CAM functionality, better cutting tools and fluids, third-party accessories, and more accurate machine tools all create synergies in a super-charged, hyper-competitive environment with very little need for regulation. I do not advocate that you should be an early adaptor of new technology, because in the machining world that is a daunting task that only a few organizations can consistently achieve. I do advocate that you stay on the front side of the technology curve. If you find yourself on the backside of the technology curve, that is also a daunting task.

LAYER 2
SKILL LEVEL

7 LEAN

What is the role of Lean in the typical machining and manufacturing organization? Each of you has likely asked yourself this question in regards to your personal responsibilities, your company, and possibly your competitors and your industry. Some of you have found the answer, but I expect that many of you continue the search. This search may be compared to a snipe hunt, finding the pot o' gold at the end of the rainbow, answering the $64,000 question, big foot, Loch Ness, or just choosing the numbers for the super lotto. But I am going to make it easy. The lean fundamentals are like apple pie and the American flag. Who can argue with eliminating waste? Who can argue with continuous improvement? And finally, who can argue with lower inventory, quicker setups, and minimal order quantities? Regardless of your size, your industry, or your personal responsibilities, these fundamentals should be incorporated.

Many Lean books, articles, and analyses focus on continuous process industries such as chemicals and foods. Others herald retailers such as Walmart and electronic assemblers such as Dell. Each industry possesses its own lean challenges and should have its own benchmarks. Inventory turns for a design and build manufacturer of specialty packaging equipment are not comparable to inventory turns at Dell or Walmart. Lessons learned in the chemical, retail, electronic assembly, and process food industry are not applicable to lean application in the world of discrete manufacturers.

Since this book targets manufacturers with machining as a significant portion of their operations, my recommendations on Lean are focused on these industries. This population of machining organizations needs to be dissected further to understand the method and extent of lean implementation. At one end of the spectrum are the auto makers and appliance manufacturers. These are high-volume operations with a predictable schedule of repeat products. It is obvious that they are practicing and benefiting from the Toyota Production System (TPS) version of Lean. For the sake of nomenclature, I am going to call these organizations the "all-Lean" companies. At the other end of the spectrum are the small companies who focus on quick-turn R&D, fixtures, gages, etc., with no expectation of repeat orders and no need to carry inventory. While these entities should wholeheartedly embrace the lean fundamentals, there is little benefit to a lean total program. I will call this group the "little-Lean companies".

In between the all-Lean and the little-Lean companies are all the rest of you that I will call the "aspiring-Lean companies". How do you know if you are in the aspiring-Lean group? If your suppliers do not have a manufacturing facility or warehouse within a few miles of your facility, you might be an aspiring-Lean company. If you have families of parts, some repeat orders, or small- to medium-batch-size lots, you might be an aspiring-Lean company. I estimate that the aspiring-Lean incorporate roughly 70% of the machining universe. Below is a list of some of the obstacles that aspiring-Lean organizations must navigate on their Lean journey:

- Outside processes: Suppliers for outside processes such as heat treat and coatings charge minimum batch fees for their ovens and plating baths. These fees cover their fixed cost of receiving, order entry, processing, certifications, shipping, etc. They will not perform these functions for a reasonable piece price for a Lean quantity on one piece. The potential EPA or NADCAP complications are prohibitive for all but high-volume manufacturers to bring these services in-house.
- Repeat orders: Minimal repeat orders or lack of knowing which orders will repeat and when they will repeat.
- Demand volatility: When repeat orders are expected, the quantity and date are unknown and both have volatility.
- Product mix: Unpredictable customer ordering leads to significant shift in product mix, leaving some equipment idle and other equipment bottlenecked.
- Customer in-sourcing: When products do eventually become higher-volume and more profitable, the customers bring the product in-house or outsource to a low-cost country despite your price reductions, excellent quality, and delivery.
- Setups: Low-volume complex components utilizing a high number of cutting tools require long setups despite reasonable efforts at setup reduction.
- Equipment utilization: Older equipment needs to be replaced with newer equipment due to maintenance, obsolescence, and productivity. Newer equipment requires higher utilization.

So if you are part of the aspiring-Lean community, which Lean tools are beneficial to your operations and how do you overcome your obstacles? You first have to realize that your situation is unique. It is easy for an outsider to lump your metalworking business into a broad category for the sake of convincing you of their knowledge of your business and their ability to effect change.

Your facility, equipment, layout, products, product mix, culture, suppliers, customers, pricing, customer ordering pattern, etc. all form a unique combination of opportunities and obstacles. I am going to call this your Lean karma. You cannot read a book or hire a consultant and implement a cookie-cutter Lean implementation. You must have a carefully crafted and customized solution that yields operational and financial improvements without forcing failure by enacting aspects of Lean that your Lean karma cannot yet accept. It is not a race or a contest. It is your business.

I am a Lean advocate. I am also a Lean realist. I have been responsible for significant Lean improvements at more than a half a dozen facilities. More importantly, I have been at the facilities long enough to live with the successes, failures, and adjustments.

None of these Lean initiatives have been easy. None of them have been linear in results or effort. None of them have been by the book or 100% Lean. All of them have been, in aggregate, successful and rewarding to the company and the people.

I believe that Lean is one of many tools in the tool box. It is an important tool and one that should be used frequently. It is an integral part of my Machining Pyramid. It is one block on the Machining Pyramid, because without the foundation and other building blocks in the Machining Pyramid your Lean efforts and sustainability will fall short. Lean concepts have been disseminated and applied for several decades. It is time that we absorb the methodology, merge the methodology with other proven concepts, adjust to the changes in the global manufacturing arena, and yield a more comprehensive and broad manufacturing management system. For machining, the system is the Machining Pyramid.

All aspiring Lean organizations should practice 5S. My strong advice is to avoid a win/lose scenario or a right/wrong scenario. 5S is a long journey. Do not make it miserable and hence negative by looking for perfection. Find areas where success is easy and then build from that success. You will invariably discover that some individual(s) will organize and actually improve productivity far greater than anticipated. You just do not know who will surpass your expectations. These workplaces inspire others, and it becomes a rising-tide-lifts-all-boats story. Your customers will be impressed with your shop, and the vast majority of your people will be appreciative. 5S will save money in the long run. There is no reason to avoid 5S and no reason to not embrace 5S.

All companies should look to create a cell where it makes sense. If you create just one cell, then you have made progress. Again, do not look for perfection and think your entire shop needs to be a series of cells. Place the components that fit into one or more cells and call the rest a "miscellaneous cell" or a "general cell" or no name at all. Do not create failures by forcing a square peg into a round hole. My tip when it comes to cellular analysis is to look for similarities and not differences. This sounds simple, but it is very important. Components do not need to have 100% of the same processes to fit into the same cell. Place some secondary equipment into the cell to handle the differences or perform a beginning or ending operation outside of the cell. You may have two families of parts that do not have similar appearance yet utilize the same equipment. Separately, they do not justify a cell. In the aggregate, they create a cell. So go and create some cells and remember, your intent is to make progress, not perfection.

You need individuals who can think outside of the box. You need some creativity. Again, if someone only views the dissimilarities between components in a family, you have the wrong team member. It is not just important for Lean analysis, but recognizing similarities wherever they occur within an organization should lead to cost and quality improvements. I cannot stress this approach, mental outlook, and personality trait enough.

Transitioning to cellular manufacturing is likely the easiest and most impactful change an organization will undertake. The improvement in quality generated by cellular manufacturing is a very simple concept. Even if you make the same quantity or frequency of mistakes in the cell as you had made in your legacy manufacturing processes, your savings will be significant, because you discover the mistake very quickly. Instead of the mistake creating dozens of scrap parts in your batch process, the cell process will "turn up" the mistake after just a few parts, or possibly after just one part. Cells may have a takt time, or more unconventional cells may process families of parts with varying cycle times and varying paths through the cell.

Structured and managed properly, the cell will simplify scheduling and capacity planning. The cell concept significantly reduces the likelihood that you will be competing for resources to meet deliveries. My experience is that the on-time deliveries for my cellular products are higher than on-time deliveries for non-cellular products. Just as important, the cellular structure lowers the stress level. All the characteristics discussed help make the cell self-sufficient and make the work simple to manage.

Other lean tenants that everyone can utilize are common management practices—elimination of waste and continuous improvement. Good leadership will create expectations and a culture that supports and rewards these efforts. All your people, at all levels of the organization should participate. Many aspects of our Machining Pyramid are intertwined with these activities. Program validation, routing integrity, root cause and corrective action, quality infrastructure—all these systematize and facilitate elimination of waste and continuous improvement.

And my personal favorite Lean tool is "push decisions to the lowest level". There is a lot more to this statement than meets the eye, ear and brain. This implies that you trust your people. It implies that your leadership is hiring intelligence and letting people think for themselves. This is how you develop people and how you empower people. Most importantly, this provides the time and focus for your technical team and operations management to perform continuous improvement, complete more projects, and transition from fire-fighting to prevention.

You may have to coach some people, mentor some people, and educate some people. But that is what good leaders do and that is what good companies accomplish. Good companies follow their ISO procedures or establish other guidelines as needed. A conscientious effort to push decisions to the lowest level reinforces the behavior. When you find yourself in a particular situation that you cannot push decisions to the lowest level, you need to ask enough "whys" until you get the answer. It is normally that the particular situation is simply out of control (equipment malfunction, too much work, etc.), there is lack of instruction/documentation, or the individual is out of position. Either way you investigate and resolve, because you will not accept a manager or technical person bogged down by routine tasks.

If you have folks in management or leadership positions involved in transactional activities, influencing routine decisions several levels lower on the organizational chart, or getting copied on copious amounts of email, then you have a problem.

Leaders who make fifty subordinates 10% more effective can move the needle. Leaders who do not push decisions to the lowest level are making the same fifty people less effective. Instead of "doing", people are asking and waiting.

WASTE

Lean practitioners teach the various types of waste within organizations and have developed acronyms to help remember. Here are the two that seem to be the most common: TIMWOOD and DOWNTIME.

T	Transport		D	Defects
I	Inventory		O	Overproduction
M	Motion		W	Waiting
W	Waiting		N	Not utilizing employee talent
O	Overproduction		T	Transportation
O	Overprocessing		I	Inventory
D	Defects		M	Motion
			E	Excess processing

I will submit that there are two additional wastes that need to be addressed. I would not consider the discovery of these two wastes to be as significant as finding two new planets in the solar system or discovering two new elements to add to the periodic table, but I would like to see lean practitioners include them into their teaching.

The first is overdesign. Design For Manufacturability (DFM) teaches us that 70% of the cost of a product is locked in during the design phase. This is easily comprehended, as components designed of very hard materials, tight tolerances, and complex profiles will certainly be more expensive than a similar component of softer material, more open tolerance, and routine profiles.

If the design engineer specifies an expensive coating, the price of the component increases. If the designer wants to be creative and re-invent the wheel instead of re-using an existing and proven design, then the price of the component or assembly increases. Existing designs possess existing supply chains, existing manufacturing processes, existing CNC programs, existing fixtures, etc.

Since 70% of the cost is driven by the design, roughly 30% is driven by the efficiency of the manufacturing process. If overprocessing is a waste, overdesign must assuredly also be a potential source of waste.

It has become common practice to outsource design and/or outsource detailing and tolerancing of the design. The outsourcing can be domestic or to lower-cost countries. Detailing is frequently performed by engineering interns and co-ops. Many of these designs are not adequately checked by another engineer with functional knowledge of the product or assembly. Our desire to reduce development costs, development lead times, and increase product differentiation fosters overdesign. Engineers perform less testing and conduct less DFM. In order to feel confident that the design will function, the designer, design team, outsourced engineer, and/or co-op will considerably over-tolerance the components. The detailer does not want to be responsible for a failure. It is faster to slap a ridiculous tolerance on a feature than to perform a tolerance stack-up, review the mating parts, look at successful previous designs, walk to assembly to review the application, and—heaven forbid—perform an analysis testing the tolerance limits. Every engineer and manufacturing personnel has seen many examples of overdesign. I have seen parts with grinding tolerances on features that hang in space with no functionality. I have seen two components that perform the identical function toleranced completely differently, more than doubling the price of the tighter component.

Compounding the overdesign waste is the rampant misapplication of Geometric Dimensioning and Tolerancing (GD&T). The design is not finished until the dimensioning and tolerancing are completed. I do not believe that you can separate the design from the application of GD&T. In many instances the GD&T is applied not in accordance to the ISO or ASTM standard. It is simply wrong. In other instances there is very little consideration given to choosing the datum structure and how the datum structure will force the part to be manufactured and inspected. Further, many locational and size dimensions are being "boxed" and held by one overall profile tolerance. This approach is simply a lazy designer taking the easiest or fastest approach to detailing a part, with no regard to the cost or complications generated by his decisions. This method may not violate the standard, but it is not the intended method of application.

The automotive industry, as you might expect due to volume, seems to be more advanced in applying GD&T to a print, with a datum structure that insures functionality but also considers gaging and manufacturing. During my time in the automotive industry I was involved in several new launches where the design engineer, quality engineer, and manufacturing engineer spent many hours in detail reviews to create datums and tolerances that met functionality requirements, yet permitted low gage error and high CPk.

The growth of corporate-supplier development personnel has more than outpaced the outsourcing of the last two decades. There is a mantra to fix the supply chain. The root cause of many delivery and quality supply chain issues are overdesign magnified by poor DFM and even poorer GD&T. These organizations need to turn inward to create and fix their prints. The bang for the buck for every dollar spent internally will be much greater than increasing supplier development staff to tell the hamsters to run faster in the wheel.

The second waste that needs to be added is borderline quality. I am not referring to defects and I am not referring to the secondary or tertiary affects generated by defects. Many machining operations generate dozens of dimensions that must be measured, adjusted, and held to the print tolerance or possibly a process tolerance. It needs to be reinforced that there are many more dimensions on a component than the few critical-to-quality (CTQ) dimensions. Just because a feature is not CTQ does not mean that the feature does not have to be held within the tolerance band. Any feature not within tolerance may cause scrap or rework. Any feature not within the tolerance band may result in a catastrophic failure for the end customer. The supply chain frequently does not know the final application and should never deviate from the print tolerance or specification, thinking that they do understand the functionality of the component or assembly.

It is not practical to sacrifice an expensive part to scrap while dialing in tooling and fixturing during a setup. If a component has three distinct machining operations, you would be losing three pieces to scrap. It is not desired to sacrifice components during the run portion of the operation due to tool wear. Any dimension created not to print tolerance generates scrap or rework. So, when the machinist is producing components that are close to being out of tolerance or were previously out of tolerance, his behavior and productivity significantly change. Below is a list of actions or a combination of the actions the machinist will undertake while fighting borderline quality:

- Slow down feeds and speeds
- Add extra passes on the cutting tools
- Add stops into the CNC program for additional inspections

- Continuously enter tool offsets
- Change tools more frequently
- Over-inspect parts during the cycle and after the cycle
- Use his indicator/probe to pick up datums and adjust work offsets
- Not run machine during breaks, lunches or between shifts

While the machinist is performing these activities, he is adding considerable time to the job. The productivity is going lower. Worse, he is not running a second machine nor is he getting ready for his next job. The result is that while there is no defect to record, the productivity and throughput is up to 50% reduced. There are times when the raw material is inexpensive, so that it is more cost-effective to ignore the borderline quality and run parts at maximum output. It can be cheaper to throw away the bad parts than to slow the machine and spend time attempting to control the process. This is not the way any process, facility, or company should operate, so it behooves the organization to identify and resolve the BQ scenarios.

How do you identify borderline quality, if there are no actual defects that create scrap or create rework that will be displayed on a metric? As discussed in other sections of the Machining Pyramid, you should have several productivity metrics that record the low output. You should also have operations management with their finger on the pulse of the shop that either detect or are informed of the issue.

I have listed some of the sources or a combination of the sources of borderline quality below:

1. Overdesign
2. Machine accuracy
3. Workholding location, rigidity, repeatability
4. Inspection methodology
5. Hard materials/excessive tool wear
6. Extended tool holders and tool deflection
7. Variability from previous operations

How do you fix borderline quality? This is likely the toughest challenge facing today's modern machining organizations. In many instances the design/GD&T has created a tolerance nightmare that is difficult to overcome. The first step is to insure you have an inspection method that is repeatable and contains the resolution required to improve the process. The intent is to always recognize the borderline-quality scenario when the machining operation is originally created. Resolving the problem is normally a joint effort between the CNC programmer, the machinists, the manufacturing engineer, and the inspector. Their reason for being is to overcome these problems, validate the program, document the setup and tooling, nail down the in-process inspection method to correlate outputs and inputs, create tool-change frequencies, etc. In these cases you will be relying on your entire Machining Pyramid. This team will need accurate machine tools, the best software and machine tool accessories, balanced holders, the best coolant-delivery method, precision-inspection methodologies, metrics, and motivated technical talent at all levels including nightshifts.

Again, the easier and the higher-volume machining has been moved to low-cost regions. Many of the remaining machining organizations competing for what is left will process new parts daily. They will create new programs and operations several times per day and at least several times per week. Problem solving and creation of robust processes with a high CPk needs to be a core competency. Your technical teams responding to a BQ issue need to behave more like a Nascar pit-crew than a Six Sigma team.

8 SIX SIGMA

The term Six Sigma has various definitions depending on the person and the organization. The most comprehensive definition and use of Six Sigma is a total management philosophy with a customer based focus to conducting business. Six Sigma is also a problem-solving methodology, utilizing the DMAIC steps (Define, Measure, Analyze, Improve, Control) that is applicable to any industry or any process. Finally, Six Sigma is a statistical measure of variation that represents a high level of quality for a specific "critical-to-quality" characteristic equivalent to 3.4 Defects Per Million Opportunities (DPMO). A process yielding a Six Sigma level of quality is the Holy Grail for the manufacturing community.

For my purposes, I will refer to the three variants of Six Sigma as Six Sigma management, Six Sigma problem solving, and Six Sigma quality level.

How do we optimally assimilate the Six Sigma variants into our modern digitally connected machining organization through our Machining Pyramid? Let us review each of the variants separately.

SIX SIGMA QUALITY LEVEL

The machining process inherently benefits from Six Sigma methodology perhaps more than any other manufacturing process. Six Sigma drives to eliminate variation. Left unchecked, a machining process breeds variation. Machining does not just breed variation at one location on one component, but on every feature on every component. Six Sigma promotes determination of critical-to-quality (CTQ) features. A typical machining process may produce twenty or more features, while only three may have been determined to be CTQ. Once the CTQ features have been determined, you should monitor and control these features with Statistical Process Control (SPC), through sampling, and/or calculation of a CPk. The non-CTQ features still need to be held to print specifications. While the CTQ features may improve the performance of the components when held close to the mean, the non-CTQ features will also scrap the component if not within specification. Therefore, we still must be concerned with non-CTQ features and control their variation to prevent out-of-tolerance or non-conforming measurements.

Let us take a brief look at why machining breeds variation. Six Sigma attempts to classify variation as either "common variation" or "special variation". Common variation is considered to be occurring naturally and will form a normal distribution. Special variation is considered assignable to a cause, such as replacing a cutting tool, and results in a shift in the mean.

Common variation in machining:

- Machine tool accuracy or repeatability: No machine tool will exactly repeat to the same position in any given axis. If you are cutting a part utilizing three axes, you have three separate interacting variances. Many positional errors on the machine will generate twice the error on the part. A simple example is turning a diameter on a lathe. A .0003" positional error will generate a .0006" error on the part.

- Tool wear: Each cutting tool is consistently wearing and not necessarily in a linear manner. The cutting-tool wear can change a dimension within a part or from part to part.

- Raw material: There may be differences within the part, from part to part, or lot to lot. These differences may cause the part to cut differently or may generate increased tool wear, which will lead to dimensional variation.

- Part temperature: Each material type has its own coefficient of thermal expansion. Ambient temperatures will generate growth or shrinkage. The metal removal process may also lead to a thermal increase. Large parts are at significantly greater risks, as temperature change drives expansion/contraction per inch.

- Machine temperature: Despite the machine tool manufacturer's attempt to maintain a stable spindle temperature during the cutting process all machine tools are negatively affected by temperature changes. An increase in RPMs will generate heat, causing machine tool spindles to expand, leading to positioning deviations in multiple axes. The machine will cool during idle periods such as between shifts, load/unload, lunch, etc., and positioning deviations will occur in the other direction. Again, if you are cutting on two sides of a feature, the positioning error will be doubled. I have measured hundreds of spindles, and they all move due to temperature changes regardless of chilling or thermal compensation. Some spindles are better than others, but they all move.

- Tool holders: Rotating tool holders contain some amount or runout. They also do not repeat their location into the spindle in an identical manner each time. The result is a variation in the tool-gage length and runout. Tool holders are also out of balance to some degree, even if you have purchased balanced tool holders. Since balance is a square function, this is more of an issue at high RPMs. Lathes or grinders have similar variation, generated by variation on tool holding, hubs, centers, or steady-rests.

- Measurement error: There will be repeatability error and reproducibility error to some degree in every measurement device or process.

- Workholding: Parts do not locate into a chuck, collet, vise, fixture, etc. the same each time. There are ways to compensate (i.e. spindle probes, tramming) but workholding will always possess some degree of variation. The best shops will utilize hardened and ground tool steels to minimize fixture movement and wear. The less of your tolerance that you give away in fixturing the better the CPk.

- Tool Deflection: Short and rigid tool holders and tooling is always better. However, your application may require an extended holder and/or extended tool. You can minimize deflection through process design, depth of cut, feed rate, RPM, etc., but you will encounter some deflection in nearly every tool, and occasionally this deflection will be excessive for the application. A CAM simulation package is ideal to determine the shortest tool and tool holder that will still provide clearance around workholding.

Special or assignable variation:

- Tool change: When a cutting tool is at or near the end of its life it should be replaced. The new tool will likely be a little larger than the worn-out tool. The length may be longer or shorter. There are several methods to measure these variables and accurately offset the difference between the programmed tool dimension and the actual tool dimension. However, there is virtually always some amount of variation in the dimensions cut on the last part with the old tool and the dimensions cut on the first part with the new tool. The reasons for this variation may be more than variation in the offset. New tools will deflect less or may contain differences in edge breaks and cutting surfaces.

- Machine tool: There will be disparities between machine tools in regards to rigidity, spindle runout, positioning, etc. This assignable variation is termed "machine-to-machine variation" when the operation is performed on more than one machine, whether simultaneously or from lot to lot.

- Fixtures: If you employ more than one fixture, there likely will be measurable differences from components machined in one fixture to components machined in other fixtures.

- Pre-existing datums: There may be dimensions or datums created in previous operations, where the variance in the earlier operations is the cause or joint cause with the variation in the current operation.

As discussed previously, I am not writing a technical machining book, but I do need to display the basic causes of machining variation to enforce that there are many sources and hence many input variables that need to be controlled to generate the desired output. When you consider that each feature on the part may have a unique cutting tool and its own set of input variables you begin to comprehend the challenge of successfully creating a component with dozens of close-tolerance features. Machining is also very susceptible to the "interaction of variables". A common example is tool deflection. Process inputs for a given part may be established that control deflection within a range that does not affect the ability to hold the tolerance within print limitations. However, within the normal variation in material hardness and tool wear the deflection is a problem when the material hardness is in the upper range and tool wear is also in the upper range.

Variation of any type is the absolute enemy in machining. The tighter the tolerance to be held, the more that variation is the enemy. Precision machining companies are climate-controlled and very clean to remove dust or small particles (from critical surfaces such as surface plates and workholding). Inspection devices and machine tools are purchased and maintained to a higher standard, just as the machinists are also trained to a higher standard. However, it may actually require more skill and effort to hold a looser tolerance on a less accurate machine tool than a tighter tolerance on a more accurate machine tool.

All machining organizations need to understand and utilize the concept of the Six Sigma quality level. It is highly advantageous for the workforce to comprehend that standard deviation is the measurement of variation, that CPk is a process capability measurement based on the variance of the data and its proximity to the mean of the tolerance, and that a Six Sigma quality level is equivalent to a CPk of 2.0 or higher.

SIX SIGMA PROBLEM SOLVING

When I evaluate the Six Sigma DMAIC problem-solving steps, I do so through the narrow perspective of machining. If you have not been exposed to DMAIC, the acronym is:

D = define

M = measure

A = analyze

I = improve

C = control

The Six Sigma community will extol the broad application and the broad processes/industries that can be addressed through the DMAIC method. The DMAIC method is generic and effective. When focusing on machining, the question becomes: What is the most effective problem-solving approach for a high number of variables and an interaction of variables all compounded by challenging metrology and precision tolerances?

Having personally been instructed and tutored by Dorian Shainin, I am partial to his teaching of statistical engineering and the Shainin System. I have enjoyed consistent success with the Shainin approach. Dorian preached "talk to the parts" and "look at the data, the parts are trying to talk to you". He labeled the dominant cause of variation the "Red X®". The dominant cause may be an interaction between two or more variables, but it is rare that the Red X® is an interaction of more than two variables. The Shainin method instructs to eliminate entire families of potential variables by determining whether the variation is occurring within the same piece, from piece to piece, time to time, or from machine to machine. This may be determined by existing data or investigation. Once this has been accomplished, you have eliminated in a very short time a large amount of input variables from potential analysis. For this reason, Shainin discouraged brainstorming and cause-and-effect approaches as slow and cluttered. With a reduced pool of Red X® candidates, Shainin provides statistical problem-solving approaches to analyze your investigation based upon empirical product knowledge. He then provides tools to confirm that you have found the Red X® and advocates being able to turn the failure "on" and "off".

The Shainin System™ highly emphasizes the measurement system. Dorian's teachings on the measurement system have been profound on my career in machining. He instructed not to move forward in your journey until a measurement system has been developed with the accuracy and resolution necessary to solve the problem. He promotes an Isoplot® study to determine measurement system variation. The Isoplot® is comparable to a gage R&R study. Dorian clearly articulated the potential of a measurement system to mask your ability to discover the Red X®. Most measurement systems, especially quality-related, have the real capability to display a marginally-pass dimension as a fail and a marginally-fail dimension as a pass. This is true whether you are employing an attribute gage such as a thread gage (go/no-go) or a variable gage such as a bore gage. Until the measurement system is corrected, analysis or investigations of any type will be compromised and your team should definitely not move forward.

The measurement system for machining is much more complex than the measurement system for other industries. Other industries measure time, temperature, pressure, or perhaps loose tolerance dimensions. The machining industry measures flatness, parallelism, concentricity, runout, location, unique profile shapes, etc. The unit of measure is normally .0001". Some companies may only need to measure in increments of .001", while other high-precision organizations require a resolution of .00001". For novices, I will provide a size analogy. A human hair is approximately .003" thick, so if you could split the hair with a razor blade into three equal strands, then one strand would measure .001". If you could subsequently split one of these strands into ten equal strands, then the ten strands would measure .0001". Many companies utilize the metric system, but the precision issues are the same. The resolution on the measurement instrument should always be at least one more decimal place than your tolerance. For example, if your tolerance is .010", you will need an instrument that can accurately measure at a resolution of .001".

There will be dozens, perhaps hundreds, of dimensions to measure on each component. Each dimension will require a calibrated instrument. Depending on size and volume, the component may be measured on a coordinate measuring machine (CMM), which normally improves speed and accuracy. Many dimensions are too small or not accessible for a CMM probe. These features will require another instrument and setup. Each of the dozens of dimensions needs to be interpreted, calculated, and measured. Any dimension can cause scrap or rework. It is not just the tight-tolerance dimensions that cause scrap and rework. Measurement accuracy is inherently an interaction between the geometry being measured, the instrument, the human, and the tolerance/GD&T requirement.

I cannot stress enough the need for those involved at any level in manufacturing to comprehend the gap in measurement complexity, cost, and skill between machining and other industries. This is especially cogent when problem solving.

I have repeatedly discovered that for machining the measurement system was often the Red X. Once the measurement error was eliminated, the input parameters provided a statistically capable process. In other words, the only problem was the variation in the measurement process. In the second scenario the measurement system was not the Red X, but the repeatability of the measurement system prevented the technical people from dialing in the input variables to create a stable process. Once the noise from the measurement system was eliminated, the correct input parameters were determined and the problem resolved. Therefore, I have learned to move quickly through the define phase of the project and evaluate the measure phase. Since resolution of the measure phase frequently leads to the Red X directly or indirectly, the project is resolved quickly. Quickness, speed, or a sense of urgency is very important in machining. If I find the RED X while the order is still running or perhaps even before the setup is complete, I can implement and verify the corrective action. This may save a tremendous amount of scrap/rework, and this permits verification while the order is still on the machine, instead of waiting until the next time the component needs to be manufactured. Most machining improvements can be "hardwired" into the process. We will either change the CNC program, alter the tool, redesign a fixture, etc. These improvements can normally be classified as a level-1 corrective action.

Six Sigma problem-solving needs to be incorporated into your organization. Since your focus is machining, the measure and analyze phases of Six Sigma DMAIC can be strengthened with the infusion of the Shainin System.

SIX SIGMA MANAGEMENT

Our modern digitally connected and data-driven factory has evolved beyond re-quiring constant projects to fix major problems. We can seamlessly prevent, detect, and repair in nearly real time. The benefits in cost and customer satisfaction are substantial. The philosophy of Six Sigma management needs to be embedded in our culture and strategic vision. The execution needs to be moved upstream in the form of corrective, preventive, and developmental actions to resolve and improve processes before they reach a pain and savings plateau requiring seal team six to fix!

If you get to the point where you need a Six Sigma project team to solve a large problem, then you should be asking if you have already failed. Paying incentives to fix problems, calculating soft vs. hard dollar savings, and dedicating teams for months to analyze long-term chronic problems does not sound like a winning man-agement philosophy. This seems like the definition of crisis management. The tal-ented people on these teams are not creating robust new processes nor are they available to resolve new problems in real time. There are certain business situa-tions that create these realities. Acquisitions, relocations, union complexities, and quality escapes with an indeterminate failure mode are the most prevalent. In these cases a Six Sigma team approach may be the only alternative.

An organizational focus on the customer and on reducing variation has become universally accepted in manufacturing and is equally beneficial in machining envi-ronments. In this perspective, the Six Sigma management methodology, teaching, and culture have been tremendously successful, as they have become mainstream thinking.

9 CNC PROGRAM VALIDATION

Our expensive capital equipment is CNC machine tools which stand for computer numerical control. As everyone knows, a computer needs basic instructions that we call a program to provide any usefulness for its owners. Without a program our CNC machine tools are just an expensive and valuable pile of electrical and mechanical components staring back at their masters. With a bad program our CNC machine tools are just an expensive and dumb pile of electrical and mechanical components. After a bad CNC program we may be able to utilize the adjective "expensive" in our description but not the adjective "valuable", as the dumb machine will wreck and destroy itself as instructed by the bad program. The machine tool certainly contains the horsepower, RPMs, velocity, and a deadly weapon in the spindle or turret to create a life-and-death battle between two or more of its own axes of motion. When this occurs, there are no winners, only losers.

We have discussed the role of CNC programmers and machinists in our People chapter. We have discussed the CAM software used by the CNC programmers in our IT chapter. And we have discussed machine tools and their programmable accessories in our Equipment chapter. The CNC program-validation process integrates these functions into a value-added exercise that establishes standards, rules, responsibilities, and a culture for creating, debugging, maintaining, and editing CNC programs. Closed-loop and Adaptive Machining techniques are further CNC programming advancements that must adhere and will benefit from the CNC program-validation process.

Each part machined may require just a few CNC programs or perhaps a dozen. Even smaller shops accumulate thousands of CNC programs that will be re-used or edited for new orders. A system for creating, debugging, documenting, storing, naming, retrieving, and interpreting the CNC programs is essential. This process separates great companies from average and mediocre companies. I preach that creating and refining the CNC program the first time is an investment, but debugging and editing the same CNC program the second and third time is a sin.

To appreciate the value of a CNC program validation process you need to understand some of the logistics and complications faced by the average machining organization related to creating and utilizing CNC programs:

1. Dozens of CNC machines of various types with diverse age, controls, accessories, and capabilities.
2. CNC programs contain hundreds, thousands, and sometimes tens of thousands of lines of code.
3. CNC programs created and edited on various shifts and weekends over a time span of years and possibly decades.
4. CNC programmers with different backgrounds, abilities, and preferences.
5. Machinists with different backgrounds, abilities, and preferences.
6. CNC programs for the same machine tool created by different CNC programmers and machinists that look and act differently.
7. Varying degrees and types of documentation.
8. Existing programs specifying cutting tools that are discontinued by the manufacturer requiring alteration to function with new tooling.
9. The same program for the same part and the same operation has to be created for multiple models of machines due to capacity and maintenance issues.
10. Machinists and/or CNC programmers performing both required and not required edits to existing programs.
11. Reprogramming due to Engineering Change Notices (ECNs)
12. Minor and major changes to the CAM software or acquisition of a different CAM package.

The premise of a CNC program validation process is simple: The machinist and the CNC programmer sign a formal document stating that the CNC program produces the specific machining sequence according to the print or process drawing. This CNC program will not be edited in the future without authorization from management. Each time the CNC program is utilized over the coming years, every machinist on every shift can be assured that the program, setup instructions, and tooling are correct. If there are complications, the problem-solving process can proceed, knowing that the CNC program is not the root cause. This is a simple statement, but the assurance that the CNC program is not all or a portion of the problem will convert problem solving from hours/days to possibly minutes. It will convert the problem solving from consuming your CNC programmer and possibly your engineer to only the machinist.

The benefits of establishing this culture are huge. High-volume producers with few programs have the capability of perfecting a program and locking the machine control to prevent editing. The high-volume organization may produce just a few programs per month. Design and build companies and contract machining organizations may create dozens of programs per day. Some of these programs will only be utilized one time and others several times over several years, but predicting which programs will repeat is normally not possible. These programs need to produce parts to prints, but it is not economical to refine them to the level of the high-volume organization. The high-volume organization is what we call "runtime-intensive", while the low-volume companies are what we call "setup-intensive". It is imperative for both to maintain the integrity of their CNC programs over the years, but it is central to the success of the low-volume companies.

Most organizations do not have a defined and rigid CNC program validation process. What they do have defined is not executed properly, and since it is not reliable, it is ignored. What are the standard operating practices at these companies? Most organizations directly or indirectly allow individual machinists to store their own version of a program either on the machine control or on a separate storage device. When a dimension is not to print, it is not uncommon for machinists to edit good programs instead of problem solving to fix the root cause. This corrupts the good program. It is common for nightshifts to change a dayshift program and vice versa. It is common for a dayshift machinist to change programs from another dayshift machinist. When receiving ECNs it is common for the CNC programmer to start over rather than edit his original program, because he has no idea what type of editing has previously occurred by other machinists or CNC programmers. This negates the power of the CAM system. It is common for the CNC programmer–when programming a similar part in a family–to start over, because once again he has no idea of what type of editing has occurred on his original programs.

Editing programs creates quality risks and editing programs saps productivity. Editing CNC programs is a non-value-added activity that creates bottlenecks in the programming department, as time is spent on edits instead of programming new parts, performing legitimate process improvements, or providing real-time technical support. Some companies have the ability to create programs on the shop floor either by installing a CAM system in the shop or through a conversational programming package installed on the machine tool. Programs created in the shop rather than in the programming department face the same type of complications to a higher degree, as there are more people involved and visibility to what is really occurring is minimal.

Why does the editing and maintenance of CNC programs default to a natural state of inefficient chaos? The majority of CNC programs are long and complex. It is typical that they will initially need to be edited to some extent, whether for functionality or for efficiency–this is a normal step in the debug and validation process. Therefore, the programmers are conditioned to making changes to their programs. They normally cannot be at the machine to watch the program operate from start to finish. The programmer needs to rely on input from the machinists. It is normally easier for the programmer to make the changes than argue–the path of least stance. This continues when the program is run for the second time. The machinists will state that either the changes were not saved from the last time or "things" are different this time.

How does an organization transition from the wild west of programming mayhem to a world class CNC programming validation process? The first step is to educate the people about the waste of editing programs again and again. Machinists, CNC programmers, engineers, and operations management need to hear that their knowledge and skill should be employed during the original program creation and debug process. They need to understand that spending the necessary time during the original creation is an investment, while the second or third time it is a waste. Just as the design of a product becomes intellectual property for the company, so should the CNC program and its associated fixturing, tooling, gages, and documentation package. Given today's technology, anyone can duplicate your design. Reproducing your manufacturing process with repeatable quality at a competitive cost should be much more difficult. Does your company view your CNC program and documentation as intellectual property?

Existing CNC programs are changed frequently by many organizations, because there are multiple machining methods and tools to create each print feature, different machinists run the same program over many years, and new tooling and/or workholding present new opportunities. A good approach is a meeting or Kaizen event to understand the different preferences, discuss pros and cons, and create standards that govern each fork in the road. This will have to be done for each type of machining. A leader needs to convey that everyone or no one will get everything they want, but what is mandatory for the good of the organization is documented standards that are followed.

All stakeholders need to understand the big picture. What is the big picture for the CNC program validation process?

1. New employees can become productive much sooner, if they have technical standards to follow and the CNC programs look and act the same. In view of the baby-boomer retirements and normal attrition, growth, and transfers, this is an important issue.
2. CNC programmers and other support personnel cannot be wasted on such a non-value task as changing existing CNC programs for preference.
3. Quality and productivity are reduced by injecting unnecessary risk into the process.
4. Constant changing of tooling utilized by a specific CNC program greatly increases the variety of tooling that must be kept in stock and reduces purchasing power.
5. The CAM system is not able to be fully exploited for productivity.
6. Human and machine capacity is consumed by the time required for unnecessary changes.

How do we balance not making changes to the CNC program with the necessity to make continuous improvement? We do want to make changes to CNC programs, but we want changes made for a legitimate business reason. There are times we know that a customer project is being phased out or the customer project is growing for the next several years. In the first scenario we would not make a change to the CNC program, but in the growing scenario we may choose to implement the same small improvement. This is the reason why your CNC program-validation process should require a designated signature to change a validated program. There has to be discipline and business logic, or there will be chaos.

There are many ways to foster continuous improvement to the specific process without editing or changing the program:

1. Refinement to the timing, direction, and magnitude of each tool wear offset.
2. Refinement to the timing of when to replace each cutting tool.
3. Refinement of frequency and method of inspection for each dimension.
4. Establishment of backup tooling that is automatically selected to replace a dull tool.
5. Performance of inspections, deburring, and tool replacements internal to cycle time.
6. Creation of a second fixture to load parts external to cycle time.

Creating the CNC program validation process is a type of work standardization. The development and execution of the standardization flushes out or reveals other weaknesses and opportunities for improvement. During the interview process I often ask perspective employees how their current or past organizations approach programming standardization. The answers are revealing and demonstrate why the specific organization has floundered or why they have demonstrated consistent success. I then ask the employee for his thoughts on how to best approach program standardization and rules for changes. The answer is revealing and helps me understand the long-term potential of the person—regardless of whether I am interviewing for a manager, engineer, programmer, or machinist.

10 ROUTING INTEGRITY

Every component manufactured or assembled should have a documented process describing the proper sequence of events. This is not only good business practice but will be required to achieve ISO or other certifications. This document is created by the manufacturing engineer with input and continuous improvement from all the functional areas.

An assembly line forces these sequences to occur in a repeatable manner every time. Machining operations can sometimes be grouped into a lean cell that creates a repeatable sequence of operations. For all the components that cannot be fitted into a cell, regardless of the reason, a router or process sheet is utilized to describe and control the sequence of manufacturing operations.

The routing is normally the controlled document as required by various ISO procedures, customers, and/or other governmental regulating agencies. This requires some level of revision control for the router. It also requires a signature and date for each operation completed along with a requirement to keep the router as part of the official "device history record".

Every assembly, fabrication, or component is assigned a unique "item number", which is also referred to as "part number" or "assembly number". Each time the assembly, fabrication, or component is manufactured, a copy of the router is created and included as part of the "production order". A unique production order number is placed on the router. A production order number is never used twice. In this manner multiple orders for the same item number will be distinguished by a different production order number, allowing for comparisons of cost and quality over time. This also provides traceability in case of a quality escape or recall. If 1000 components are manufactured during a two-year period, you might be subject to a recall of all 1000 components, if you cannot discriminate between them. If you produced these components on ten different production orders and can determine which production order yielded the faulty component, you will only have to recall and replace 100 of the components.

So why is "routing integrity" an integral block in our Machining Pyramid? The answer lies in the myriad of information above and beyond the process steps contained within the router and how this treasure of information is utilized within an ERP/MRP system.

A router displays each internal and external manufacturing operation needed to create the component from raw material to finished part. In the case of an assembly, the router will also list each operation sequence, including any testing or inspection procedures. These operations are listed in the proper sequence. Each type of machine tool or outside process is assigned a unique alpha-numeric designation that is usually referred to as the workcenter number. Hence, a small vertical machining center may be VMC001, a small horizontal machining center may be HMC001, and so forth. Each type of machine, deburr, clean, assemble, fabricate, or weld station is assigned a unique workcenter number. Experience and logic is helpful when establishing workcenter numbers, as the workcenters establish scheduling and capacity management.

The ERP/MRP scheduling module will create a schedule or dispatch list for each workcenter. If the average shop completes six hundred operations per week, it is imperative that the workcenter schedule is meaningful and followed.

The router should also contain the expected setup time for each machining operation and the "runtime" per piece for each operation. If the setup time is insignificant, it can be omitted. A general rule is that these times are accurate within 20 % for low-volume components, 10 % for medium or batch volume, and within 5 % for higher volume. Since your ERP/MRP system is loaded with the number of people and number of shifts available for any given workcenter, the dispatch list can calculate the number of hours or days required for an operation to be completed at any given workcenter. This also supplies accurate data for capacity management. I will discuss scheduling and capacity management more thoroughly in future chapters.

Another important aspect of the router is the bill of material (BOM). The router will dictate which operation requires raw material or any sub-level components for any type of assembly or weld operations. This relationship creates the functionality for ERP/MRP to generate demand and schedules for all sub-level components when a demand for the assembly is created. The assembly is sometimes referred to as the "parent", and the assembly sub-level components are frequently referred to as the "children".

While the BOM is sometimes a separate module in many ERP/MRP systems, it is the routing that dictates when the children are required, if they are backflushed, how many pieces, inches, or pounds are required, etc. If this data is not correct, then inventory quantities will be wrong.

An important aspect of routing integrity is the contribution to building a quality product. In the routing for each operation the engineer can provide instructions concerning what and how the work should be performed. This can include information about lessons learned from previous production runs. This information is particularly helpful for the nightshifts, weekends, and new employees. It is just one way to capture the tribal knowledge from experienced employees and to benefit from past experiences. This is especially helpful in low-volume/high-variety organizations where components do not repeat for months or years.

The routing instructions should also clearly specify which print features are being completed at each operation. More complex components may require features to be "roughed" at one process operation and finished at a subsequent operation. Many components will be roughed prior to heat treat or will be machined to a "process" dimension to allow for an expected coating thickness. While some of this detail may be in the CNC programs and CNC setup instructions, it is common for the CNC programmers to utilize the router as their work instructions to create the process.

This Machining Pyramid building block is named "routing integrity". I want to stress the word "integrity". In most cases, organizations are using some type of routing system. In most cases these organizations cannot use "integrity" to describe the accuracy or usage of their routers and hence their production orders. If operations are performed at workcenters not designated for production orders, the effectivity of the scheduling system is reduced. Any ERP/MRP will create a schedule based upon the workcenter on the router. Deviation from the router creates scheduling fluctuations in both the workcenter being skipped and the new workcenter being added. More importantly, there will be a higher risk for a quality issue and probably higher manufacturing costs. The alternative workcenter may require a new CNC program, new tooling, new workholding, and new work instructions. All of these involve quality and productivity risks.

If the routers and corresponding production orders do not contain accurate setup and run times, the scheduling system will be compromised. Again, ERP/MRP can only schedule based upon the times provided within the router for each operation. ERP systems utilize the setup and runtimes for standard cost, estimated cost, or both. Inaccurate times and inaccurate workcenters will lead to inaccurate standard cost and poor schedules.

Labor reporting utilizes the setup and runtime to compare the actual performance to determine efficiency, which provides a daily progress report, if the production order is on track for creating profit or creating a loss. The efficiency metric also provides comparable analysis between different shifts and different machinists. Part of routing integrity is to review these metrics and adjust the routers so that they are more accurate for the next production release. A corporate objective to achieve routing integrity is a component of continuous improvement. Responding to anomalies on the labor reports will help insure that the sequence of operations, the workcenters, time standards, instructions, and notes on the routers are optimal.

The production order contains the latest engineering revision level from the customer. Each person in the organization who is involved with the order should verify that the engineering revision level on their print matches the production order. Each CNC program and quality document should also contain the latest engineering revision level and be verified prior to use. It is not uncommon to produce multiple engineering revision levels of the same part, whether for the same customer or for different customers. It is also not uncommon for older documents for the older engineering revision levels to accidentally be utilized.

There are times when a NCMR disposition mandates that components need to be reworked. In these cases, the manufacturing engineer will obtain any required approvals and subsequently add the required operations to the production order with the necessary instructions. The ERP/MRP can then schedule these new operations. Since the production order is part of the device history record, this process meets all ISO requirements.

A routing system with detailed instructions, accurate time standards, and revision control is simply a primary driver for quality, productivity, and on-time deliveries. When there is an experienced and dominating team in professional sports, it is often said that the road to the World Series or the road to the Super Bowl runs through XYZ, meaning that any team wanting to go to the Super Bowl must defeat the New England Patriots on their home field. In machining, the road to the Super Bowl runs through the New England Routers! The routers control the cost of the product, the expected absorption of overhead, scheduling, capacity management, labor reporting, quality instructions, subcontracting sources, and inventory accuracy through the BOM. The router is an ISO-controlled document, contains the process revision history, becomes the production order and the legal device history record with signatures and certifications. Most companies encounter fundamental problems that last for years where the root cause relates to routing integrity.

LAYER 3
EXECUTION LEVEL

11 SCHEDULING/CAPACITY MANAGEMENT

Scheduling and lot size should not be an afterthought left to production control and manufacturing. Scheduling methodology along with the purchased and manufactured lot quantities should be an overall business strategy that is supported by a sales and operations management strategy. Reducing lot sizes requires sales to establish repeat customers, who order repeat or similar products with a rhythm to the volume and with predictable lead times. With this accomplished, operations can install the appropriate equipment either in a cell or not in a cell. Fixtures, tooling, and processes can be created to minimize setup times and establish the necessary cycle or takt time. In other words, lot sizes need to be "engineered"—not random, forced, or dictated by an executive or consultant. An example would be transforming hundreds of discrete parts, each with its own routing, into a small group of part families with a machining cell for each family. Each cell would reduce setup time and lead time while standardizing tooling, workholding, and metrology for the part family. This would permit smaller order quantities to be processed.

Another example would be transferring a part family with high-batch production from a vertical machining center with a tool capacity of twenty to a two-pallet horizontal machining center with a tool capacity of one hundred. Each side of the four-sided tombstone may be configured for a unique part number. Eight distinct part numbers could be left with the setup maintained on the two tombstones with zero changeover as the result. All tools would fit in the one-hundred tool carousel. Additional tombstones, pallets, or tool positions could be added if needed. These engineered solutions for setup reduction, lot size, and scheduling simplicity are predicated on the sales strategy (customer stability, volume, delivery plan), as investment in the wrong parts families or customers will be expensive and prohibit or restrict equipment from other orders. Do not expect manufacturing to work miracles one order at a time without alignment of strategy with sales and engineering.

Before I discuss the pros and cons of scheduling methodology and how to manage the entire scheduling system, I want to review the various types of machining organizations and their associated scheduling complexities.

- Original equipment manufacturer (OEM) with captured production. These organizations fall into the design and build category that has been discussed throughout this book. Examples include machine tool manufacturers, custom machinery builders, and suppliers of automation equipment. These organizations generally will be producing low-volume orders with a high variety of part numbers. Scheduling complexity is high, especially considering that there will be a need to purchase items from the supply chain and to manufacture non-forecasted spare parts for the field. Purchased items will vary from commodities, purchase to print, purchase partial manufacturing of a component, and purchasing of processes such as heat treat and coatings. The challenge for the OEM is to create the design, manufacture, assemble and ship in a specified time frame. There are many potential pitfalls in this process that will create late deliveries, scrap, or margin loss. Chief among these pitfalls is debugging a new design and a new manufacturing process for each new assembly and part number. As it relates to scheduling, this is a complex task to sequence delivery of all levels (raw materials, purchased components, manufactured components, sub-assemblies, assemblies, parent) of a bill of materials (BOM) to facilitate build, test, run-off, and installation.
- Job shop: These machining organizations are normally the smallest of the machining companies yet may have high flexibility and high individual talent levels. The owner is normally the entrepreneur who drives sales, quoting, and daily operations. The type of work is frequently R&D-related or low-volume make-to-print for OEMs. The vast majority of orders do not repeat. Job shops compete on flexibility, creativity, and short lead times. It is not uncommon for a single tool maker to complete all necessary operations. Scheduling complexity will vary widely, depending on the size of the shop and the amount of work at any given moment. Many of these companies do not employ a formal scheduling system, while others utilize scheduling software or ERP/MRP.
- Contract machine shop (CMS): This type of organization will ideally obtain either a blanket purchase order, long-term agreement (LTA) or a vendor-managed inventory (VMI) contract. Contract machine shops may range in size from a dozen employees to more than a thousand. The contract machine shops generally possess just a few niches of expertise, which can be leveraged to a large customer or many customers within a specific industry such as medical or aerospace. The contract machine shop is prone to possess much lower overhead than its large customers yet demonstrates high agility and expertise in its specific niches. It is normal for the CMS to work directly with the customer's design group to nurture products from concept to production. The ability to provide Design for Manufacturability (DFM) advice along with early-stage R&D machined components directly leads to multi-year contracts. Lot sizes range from single digits to hundreds of pieces. It is typical for the CMS to utilize ERP and manage hundreds and

possibly thousands of production orders at any given moment. Many CMS will produce a high number of R&D or experimental orders for each part that eventually matures into an LTA or VMI. Low-volume parts naturally require more operations and workcenters, which exacerbates the scheduling.

- High-volume manufacturers: These organizations include automotive, appliance, fasteners, and other commodities. These industries utilize dedicated equipment with very little change-over. While the contract machine shop, job shop, and OEM are typically setup-intensive, the high-volume manufacturers are strictly runtime-intensive. The high-volume manufacturers strive to remove seconds from the processing time, while the setup-intensive organizations would not ever consider investing effort to remove seconds. High volume also justifies development of custom raw material (investment casting, extrusion, forging, etc.), equipment, workholding, and tooling. All of this special engineering of the machining processes will reduce the number of distinct operations and workcenters required to machine the component which drives down cost but also simplifies scheduling.

A ranking of the four types of machining organizations according to scheduling complexity with the most difficult first would yield the following result:

1. Original equipment manufacturer
2. Contract machine shop
3. Machining job shop
4. High-volume manufacturer

The obvious questions are what type of scheduling system to utilize for your organization, and what are the secrets to success? When discussing scheduling methodology, the first question is always whether to "pull" or not to "pull". Lean advocates, with religious fervor, will always recommend a pull system. The drive and enthusiasm displayed by lean practitioners is admirable. Every company would benefit, if this level of energy and commitment were displayed throughout the organization. I suspect that the Lean advocates will only be embalmed at a funeral home where the mortician has a one-day supply of embalming fluid and processes the body in a U-shaped cell.

There are certain applications for which a pull system is optimal. There are many applications for which a conventional pull system is not optimal but with modifications will be ideal. Again, some Lean advocates insist on no modifications to the text-book pull system, but I do not agree. I believe that the difference is the perspective of academia or a consultant versus a plant manager who has to actually execute through real-life scenarios that play out over many years. Land may be cheap in a flood zone and the view may be beautiful, but is it the best place to build your home? Everything can be fine until the flood arrives. Let us take a few minutes and review the traditional Kanban pull system along with the benefits and pitfalls.

Kanban began at Toyota after WW II out of necessity due to the lack of capital, facilities, and presses. Toyota realized that small batches would permit their capital equipment to run many different part numbers. They discovered that small lot sizes actually prevented large quality issues and found ways to reduce setup time (Single Minute Exchange of Dies, SMED), thus permitting even smaller lot sizes. Instead of a single large press devoted to one vehicle body part, a single press could create many body parts.

As components were assembled to meet customer demand, Toyota "pulled" replacement parts from the preceding operation. Kanban cards were used to signal to the preceding operations what and how much to manufacture. Even when Toyota became successful and possessed the capital for new equipment and facilities they did not revert to Western manufacturing methods of large-batch production and large inventories. Toyota used their knowledge and resources to refine the Toyota production system with smaller hydraulic presses and more efficient cells.

The success and simplicity of the Toyota production system has spread within the automotive sector and to other industries. In an environment similar to a Toyota production facility, and the pull system set up and managed correctly, the results will be a simplistic, effective, and visual scheduling system. If you do not have a sixty-day locked schedule and your suppliers do not have warehouses full of your product adjacent to your facility, then your pull system may be just a bit more challenging.

CHARACTERISTICS OF A GOOD PULL ENVIRONMENT

What is the right environment and application for a Kanban pull system to flourish? In our Lean chapter of the Machining Pyramid, I described three types of companies in the Lean spectrum: all-Lean, aspiring-Lean, and little-Lean. As a refresher, the all-Lean were the high-volume manufacturers such as automotive and appliance. The aspiring-Lean were the original equipment manufacturers and contract machine shops, while the little-Lean companies were the machine job shops with very little repetitive orders (and no rhythm to the orders that do repeat). Utilizing a pull system for scheduling follows a similar path. The all-Lean companies possess a suitable foundation for pull, while the little-Lean do not. The aspiring-Lean companies may have a suitable environment, or they may be able to adapt/modify their environment and/or, heaven forbid, adapt/modify the pull methodology to their environment.

Good pull characteristics:

1. A "rhythm" to the customer demand! Whether four per month or four thousand per month, there needs to be reasonable predictability to fill the pipeline and establish pull quantities for the internal and external supply chain.

2. Cellular manufacturing is much more conducive to a Kanban pull system than traditional manufacturing. The reduced travel distance and all/most operations within line of site and close proximity establishes ownership into a small group of people. Cellular manufacturing also implies short setup times and cross-trained people that can perform all operations within the cell.

3. Components or assemblies with no processes to be performed by outside suppliers permit full control of all transportation and manufacturing operations to remain in-house, thus providing predictable lead times and minimum risk. One outside process that is very predictable can still be workable with a pull system. More than one outside process or any unpredictable lead-time or quality issues and the risk of a stock-out situation occurring is too great.

4. Excess machining capacity at each specific operation is required to prevent stock-out situations. With little or no buffer in the system, demand volatility or supply disruptions will lead to late deliveries. Supply disruptions may be equipment failure, supply chain delays, scrap/rework, or personnel issues. Demand volatility may be as simple as product mix changes from multiple customers, creating higher demand for product "D" and lower demand for products A, B, C, E, F or G. Once behind, a pull system has very poor visibility for prioritizing late orders. Excess capacity allows quick recovery and minimizes the human input required to set scheduling priorities for long periods of time.

5. Assembly operations and simple fabrication processes. This type of manufacturing is predictable, requires less space, less setup, less measurement, etc.

COMPANY L

So what are the consequences of utilizing a Kanban pull system in an environment not properly suited for pull, also known as a square peg in a round hole. I will describe a nightmare from a personal experience. I inherited a medium-sized facility that machined, welded, and assembled valves and couplings for the oil and chemical industries. I will call this organization "Company L".

All production and purchasing was a two-bin Kanban pull system. There were eight different product families manufactured, with each one in its own lean cell. Each product was sold in varying diameters, lengths, material type and seal material, all depending on the application. This product variation resulted in approximately two thousand pull cards. Each assembly station was provided an assembly schedule each morning that contained the orders left over from the previous day's list and the new orders sold within the last 24 hours. When the assembler emptied a parts bin he placed the pull card in a pre-specified visible location. Purchasing collected the purchase cards each day, and the cell machinists collected the cards for the machined components. Each machine was color-coded, so the machinist placed each card on the color-coded Kanban board next to each machine. Each pull card contained the bin quantity, which instructed the machinists how many parts to manufacture.

I made some immediate observations that I considered to be red flags:

1. The assemblers frequently would build a partial customer order due to lack of parts and were not able to ship the order. Example: three out of five or five out of six units completed.

2. Due to size and variation the assembler needed to walk to collect all the components on the BOM for an order. He would be missing one component and not be able to begin to assemble the order. Instead of putting the remaining components back into the bins, the assembler would place all parts on a shelf with the missing part(s) highlighted on a parts list. When I enquired why this was done the answer was, "if I put all the parts back, then tomorrow when I pull them I may be missing a different part. This way I know what I'm missing, and when it arrives I can continue the order".

3. The Kanban board at each machine tool had three hooks and a pocket for overflow pull cards. The problem was that there could be 15–30 cards at any one machine tool. This could be 3–4 weeks of work even if no new cards arrived. It was not uncommon to have both pull cards for the same part number waiting in queue at the machine tool.

4. The pull cards frequently had a piece of tape or a post-it-note over the quantity, with a new quantity written on the tape or note. The new quantity was normally greater but occasionally lower.

5. The machinist would sometimes not have the forgings or raw material necessary to make the required parts as instructed by the pull cards.

It is obvious that these red flags generated inefficiencies, late deliveries, and headaches. First, the components sitting idle at the assembly station, whether assembled or not, assembled locked-up parts that could have been used for other orders if they had still been in their bins. Second, it was nearly impossible to determine the priority required to run the pull cards at each machine tool. With extensive effort, the supervisor attempted to review the assembly schedule to ascertain total demand for each separate part number and then count the quantity of each part number that was still in the bin as well as already pulled out of the bin by the assembler but sitting idle waiting for the remaining part(s) to arrive. This required memorization of the BOM for each assembly or reviewing the BOM for each order. This exercise was repeated for each card sitting at the machine tool to manually reconcile available inventory and demand at assembly to establish priority of all the pull cards sitting at the machine tools. This highly manual and time-consuming activity had to be repeated each day. Yes, it seemed like Ground Hog Day without the laughter. The tape and notes over the quantity were a result of the supervisor totaling the demand from the assembly list and noticing that the Kanban pull card quantity was considerably different from the demand. This could occur for two reasons. First, a customer may place a large order for new construction or to perform PM during a scheduled facility shut-down. Second, random orders from multiple customers during a similar time period aggregated into a larger demand than the pull system had established. Either of these conditions could consume one or both bins of stock. There was no mechanism in the Kanban pull system to recognize the higher demand.

I made several changes that yielded marginal improvements. Despite a union shop I temporarily subcontracted to reduce the machining bottlenecks. The organization did not want to invest in a union shop, but with considerable persuasion reluctantly agreed to add equipment. Next, new cards were printed with new Kanban quantities based on a two-year order history.

With multiple Kaizen events and many changes in the Kanban pull system Company L improved but remained inefficient and complex. It is a fair question to ask whether the remaining problems at Company L are management-related or cultural, or whether it is just not a good environment for a Kanban pull system. I suspect a little of all. The pull system was not scalable as the company grew. Growth can be in volume and variety. I find that variety is always a more challenging logistic. The manual pull system had no access to electronic data and provided no transparency. Physically replacing the Kanban cards with updated information was cumbersome due to the quantity, color, lamination, and analysis. Needless to say that this was performed very infrequently—maybe every 4–5 years. The market created product mix changes faster than the cards were updated. There was no mechanism to respond to firm customer orders with demand that totaled much higher than the Kanban quantities. The result was that a machine tool could produce the Kanban quantity, and the assembler would empty the bin the same day with demand yet to be filled. Similarly, purchasing ordered the Kanban quantity, which was frequently insufficient for existing sales orders.

There were no stocking locations or inventory locations within the ERP system. The culture was to use the KPS regardless of how much manual intervention needed to be spent to apply band aids that were inefficient. Was there a better way for Company L to operate? Yes, my solution for organizations similar to Company L will be discussed later in this chapter.

CHARACTERISTICS OF A BAD PULL ENVIRONMENT

I have listed environments that are favorable to a Kanban pull system, so here is a list of characteristics that are not favorable to a Kanban pull system.

1. Customer demand that is infrequent and unpredictable.
2. Raw material or sub-level components that are not in stock at your location, not available in a few days, or have an unpredictable delivery lead time from your supplier(s).

3. Assemblies or components that require design or process engineering prior to the start of manufacturing. Industry examples include machine tool manufacturers, custom machinery builders, and suppliers of automation equipment.

4. Complex machined components that require multiple setups, multiple machine tools, and/or a high number of cutting tools and inspection features.

5. Machined components that require outside processes, especially multiple outside processes with minimum lot charges and certification requirements.

I expect there will still be some Lean advocates who would suggest that all components can be processed via pure pull. There are more than 30,000 companies performing machining in the U.S., and a similar number in Europe. I would contend that many companies must quote orders and do not know whether they will receive the order, and when they are successful and receive the order, they do not know if they will ever receive an order to manufacture the component for a second time. The majority of these items require many setups and a number of outside processes. These will never be on a pure Kanban pull system. It is not practical for most companies to be highly vertically integrated to eliminate outside processes. Most aerospace, military, and medical OEMs require heat treaters, coaters, and testing companies to be approved suppliers to the OEM and to maintain NADCAP approval of each special process. As a supplier of machined components to the OEM you are directed where to sub-contract each specialty process. These organizations require minimum quantities and a minimum lot charge. Most of these special processes require an initial multi-stage cleaning before the special process is performed, and testing after the special process is performed. All these activities are batch-related. A destructive test may be required to validate the lot. They are not going to dip a few parts in several tanks for one hour each without charging you the full amount. These processing companies have a high fixed cost to enter your order, insure compliance to all specifications, perform the testing, and create the certification documents.

AS9100 requires full first-article inspection, if the assembly or component has not been manufactured at your company within 24 months.

The spectrum of readers comprises organizations who clearly possess an environment suitable for a Kanban pull system and those whose environments do not. Some of the pull environments have moved to low-cost countries, as this includes the repetitive and simpler products. Other organizations have developed a pull environment as a means to remain competitive and to remain in a high-cost region. A majority of the machining organizations remaining in high-cost areas are those with complex, low-volume, and design content that do not have an environment suited or adaptable for conventional pull (see Figure 11.1). Fortunately, there are two methods I will discuss that provide scheduling systems that incorporate the lean philosophy but accommodate the realities and complications of the low-volume environment. Advanced Planning Software (APS) is installed in addition to your ERP and is more prevalent for OEMs. For contract manufacturers your existing ERP system can also be combined with traditional pull to create what I have called hybrid electronic pull.

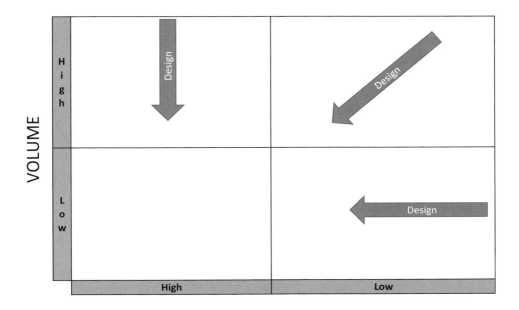

COMPLEXITY

Figure 11.1

HYBRID ELECTRONIC PULL SYSTEM

Technology changes everything! You only have to look at the phone in your hand. Your phone is also a watch, a book, a calculator, a newspaper, a flashlight, a camera, a video game, and many other devices. What did your phone look like when Western manufacturing began to adopt lean principles and implement Lean Kanban pull systems? Your phone will most probably still have been attached to your wall! Automate, emigrate, or evaporate! Machining must embrace modeling, simulation, digitization, electronic tool setters, spindle probes, and much more. We must also apply technological advances to the application of Lean. Academia and consulting professionals want to be purists and typically espouse no changes to the traditional Kanban pull system. Utilizing technology and adapting new techniques while maintaining the underlying principles of Lean is the real respect and evolution of the philosophy. Should we not apply continuous improvement to Lean itself?

If your profession is running a manufacturing organization, your job is to save every person every minute possible both in the factory and in the office. More importantly, you need to reduce frustration and waste including those associated with the constant prioritization of pull cards.

The answer is to eliminate the pull cards by replacing them with an electronic card. Electronic pull signals (EPS) can be displayed at each work area on a simple PC or a widescreen TV. The quantity on an EPS can be updated instantly. Most importantly, the priority of an order on EPS can change in real time. If every order on the EPS list at a machine tool is not needed for two days, and the raw material arrives at the receiving dock at 10:00 a.m. for an order needed by the customer today, the EPS for the order just arriving can go to the top of the list at 10:01 a.m., notifying the machinist that the raw material has arrived.

Utilizing the EPS approach widens the application of pull, since it allows many cells and facilities the agility to operate efficiently that otherwise could not when restrained by a manual pull system.

Toyota and other automotive companies lock in schedules and demand weeks and months in advance. The preponderance of machining companies do not have this luxury. The average machining company must respond to customers who change quantities and dates, while also responding to surge demand due to multiple customers unexpectedly ordering the same item or similar items requiring the same workcenters.

Every ERP system tracks demand, inventory, and WIP. When new demand arrives, whether an external source or internal source (assembly, scrap), a new EPS will be created or an existing EPS may be increased. At the same time the EPS will be prioritized and sequenced in relation to other demands. The prioritization can constantly occur electronically, thus avoiding relentless human intervention throughout the shop and permitting the support personnel to perform continuous improvement rather than continuous manual non-value-added pull card and bin shuffling.

So what are the basic logistics behind successful use of a hybrid electronic pull system (see Figure 11.2). You probably already possess the software and tools necessary for implementation. Whether you currently practice Kanban or another type of scheduling, you possess some level of ERP package to enter sales orders, write POs, invoice, schedule, etc. Newer cloud-based ERP systems are ideal for smaller organizations, as the cost is lower, IT demands are less, and the implementation is faster. The MRP dispatch list from any system can be adapted or modified to serve as the electronic signal. This electronic signal can be established per workcenter or per cell.

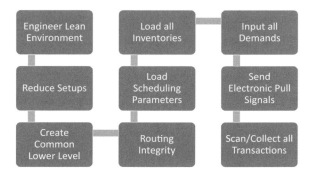

Figure 11.2 Establish Electronic Pull System

STEP 1: Your physical layout and your processes should follow lean principles. Where your environment is suitable you should employ cellular manufacturing. Processes need to be standardized and robust, and waste needs to be eliminated.

STEP 2: Reduce setup time. Lot sizes can be smaller when fixed costs are reduced. Reducing setup time leads to better efficiencies, but the effort and teamwork required to reduce setup times also generates simpler methods, better quality, and less time and product variation.

STEP 3: Many part families may share sub-level components, or different parts in a part family may be identical until differentiated at the end of the process. Each part and family should be engineered and analyzed to create common raw material or partial components that can be processed in small groups as far towards completion as possible, awaiting a customer order. They can then be pulled electronically the rest of the way through the process and shipped quickly when the customer order arrives.

STEP 4: Your routers must contain each manufacturing sequence by workcenter and include a setup and runtime. An entire cell may be one workcenter, if you choose. The BOM needs to be tied to the router and all lower-level components are tied to the appropriate sequence. All outside processes are included on the router with the embedded queue time necessary to complete the process and return shipments.

STEP 5: Load your scheduling system with the number of shifts and hours available at each workcenter. Your scheduling system now has all data necessary to schedule for each workcenter to achieve the required delivery date.

STEP 6: All inventories (finished, WIP, raw) must be accurate and entered into ERP to include quantity and location.

STEP 7: All demands need to be entered electronically, including due dates and quantities.

STEP 8: Depending on your business model, the required component(s) or assembly to fulfill the order may currently be in finished goods inventory, WIP, raw material, or non-existing. LTAs, VMIs, and warranties may dictate a certain level of finished goods. The hybrid electronic pull system will identify the location of the needed component(s) and send an EPS. If that location is finished goods, then the component(s) will be pulled and shipped to the customer. If the location is not finished goods, an EPS will be sent to the HEPS screen for all remaining operations. Each workcenter should contain two HEPS lists. The first is the "active" list, which signifies that the order is currently at the workcenter and available to run. The second is the "standby", which signals that the order is needed but not yet available. As the order is pulled through the system by the demand, it will immediately and automatically move from the "standby" to the "active" HEPS screen. In all cases, the priority and quantity will be automatic. Any customer change to scheduling can be reflected at the workcenter either after MRP cycles or instantly with a production control update. Since we are responding to actual customer orders through MRP, we are not making for stock and not pushing—we are pulling orders through to fulfill a customer demand.

STEP 9: This hybrid scheduling system requires people to immediately close each sequence on the production order when completed. The number of good pieces and the number of any scrap pieces are both recorded. This process can be automated and also serves to provide real-time tracking of orders for production control or sales as well as a record of the time required to complete the sequence. Traditional KPS do not provide this tracking or instant visibility to scrap. This is not significant for simple components but is highly advantageous for components with many operations and longer lead times.

Taiichi Ohno originally transferred the Kanban philosophy from observing American grocery stores. Many companies refer to their stock of parts with their Kanban system as their "supermarket". Technology has changed the American grocery stores during the last three decades. Large chains who have become proficient (Walmart, Kroger) at integrating traditional grocer and retail practices with technology to create operational excellence and marketing knowledge have dominated the industry. Grocers who failed to adequately automate have disappeared or fallen behind. No longer is there simple visual control of shelves, clip board checklist, stock-boys replacing sold items, and re-ordering based on empty space. The modern grocers, large and small, utilize scanning to automatically replace inventory throughout the supply chain. This technology supplies big data about individual customer purchasing preferences and provides customized discounts and marketing to maximize customer revenue and loyalty.

The evolution of material flow, scheduling, logistics, and technology in the grocery industry that was the spark of Kanban is the same evolution that needs to occur in manufacturing industries. You should utilize bar coding to replace your inventory and gather the data to determine quality metrics, productivity metrics, current margins, and future pricing. Grocery stores in the second decade of the 21st century are not the same as your parents' grocery store or the store of even your youth. Looking back, I recall two types of milk, one type of orange juice, and a handful of cereal brands from which to choose. Organic and diet products did not yet exist. I counted thirty-three different types and sizes of milk and sixteen varieties of orange juice on my last trip to the locally owned mid-size grocer. This explosion of variety witnessed at the retail level is mirrored by the mass customization phenomena in manufacturing. Your manufacturing shop should not be your father's shop. Embrace the technology and software that empowers logistical mastery of today's lower-volume, lower-inventory, and higher-variety world.

Consider the following changes in market conditions and their effect on Western manufacturing facilities.

- Simple production and high-volume production have moved to low-cost countries eliminating facilities with simple logistics and simple scheduling.
- Mass production has been replaced by mass customization. Thirty-six types of milk and thirty-six different designs to be processed and machined instead of one.
- Assembly lines and machining transfer lines replaced by cellular manufacturing and CNC machine tools.
- Customers who now demand one shipment a week (or day) for a given part number versus one shipment per month or quarter for the same part number in the past.
- Customers who were ecstatic with 85% on-time delivery are now upset with 95% on-time delivery.
- Customers who issued blanket orders once per year now provide access to internet portals, where schedules fluctuate weekly for both quantity and date.
- Solid models that designers can clone and produce a new iteration in minutes which dictates that manufacturing must also follow suit.
- Inexpensive and advanced reverse engineering technology that permits competitors to copy designs as soon as they hit the market. These may be legitimate fast followers, or they may be stealing patented information and selling on the black market.
- Cyber hackers stealing intellectual property, including research, designs, marketing analysis, strategies, and proprietary manufacturing processes before new products even hit the market.
- Employee migration to competitors and across borders, transferring design and intellectual property.
- Sharing intellectual property with a partner and future competitor as a reward for spending millions on new facilities, equipment, and training.

These trends all dictate shorter product lifecycles, reduced development schedules, shorter lead times, higher product differentiation, and lower volumes. Profits are reduced rapidly as competitors release similar or improved items at a lower cost. Corporate strategy has focused on being the first to market and to rapidly update products to stay ahead of competitors. This strategy applies to business sectors ranging from medical over transportation and to energy. Why settle for beating the competition when you can eliminate the competition by being the first to market. What does it mean for the average machining supplier? It signifies much more difficult logistics and scheduling. To survive and thrive requires more sophistication than just good machining skills.

ADVANCED PLANNING SOFTWARE

APS provides scheduling simulation along with advanced graphics that create agility for manufacturers to establish delivery dates and adjust to scheduling disruptions. This software replaces the existing MRP package but relies on the Item Master, BOM, and routing from your ERP. OEMs require multi-level BOMs that require a complex supply chain of internal and external sources. It is common for some of the BOM items to be in stock, some to contain a short lead time, and some to possess a longer lead time, whether originating internally, domestically, or internationally. The OEM must supply commit dates to customers during the quote process, at order acceptance, and throughout the build cycle. The APS tools help determine whether delivery dates can be achieved without disrupting other orders. The APS graphics powerfully display the entire BOM but highlight the items restricting delivery. The scheduler can make the necessary adjustments to achieve the required delivery date or ascertain the new date, if there is no solution to correcting the project.

Purchased components from international sources and components with new design content are frequently the longest lead time items. APS allows the scheduler to quickly move the entire assembly to coincide with the last arriving component, thus creating capacity for projects capable of being completed at an earlier date. The entire assembly or project may be rescheduled to a new date with a simple click and drag of the bar to a new date on the chart.

The APS will require fulltime a scheduler(s) to respond to any changes in supplier deliveries, equipment downtime, scrap, etc. The software is very user-friendly and permits very fast simulation and rescheduling. As the Machining Pyramid displays, the APS scheduling can only be effective with accurate routers, labor times, data collection, and BOMs. Purchasing or supply chain management must input supplier acknowledgements and any supplier updates.

There is a subtle management difference between the conventional ERP scheduling used for the hybrid pull system and an APS approach. An ERP scheduling system is normally based on infinite capacity and relies on humans not to overload the internal or external supply chain. If there is a moderate overload, the conventional ERP relies on operations management to assign overtime or move personnel to compensate. An APS is normally managed in a finite scheduling approach, and the scheduler informs operations management where there will be scheduling over/under-capacity and how to respond.

SCHEDULING – LOT QUANTITY

Regardless of your scheduling system, a controversial decision is always the lot quantity, which is also referred to as order quantity or pull quantity. Simple and inexpensive components with constant demand are an easy lot-quantity decision. Expensive and long lead-time components with intermittent demand can be more challenging. There are universal principles to apply that assist this decision, re-gardless of where the component lies on the spectrum. Always continue to drive lean initiatives to reduce setup times and overall lead times. However, when mak-ing decisions on lot quantities you have to accept that "you are where you are" on your lean journey. If you can only bench-press 100 pounds, you may be in trouble if you put 200 pounds on the bar. You will be in trouble just the same if you are too aggressive and lean too far over your Lean skis. You need to insure customer deliv-ery first and not choose a lot quantity that is either too large and consumes current capacity, or is too small and consumes current capacity through multiple setups. The scheduling priorities are demonstrated in a Pareto format in Figure 11.3. Since our focus is on the customer, the Pareto shows the following sequence: Delivery,

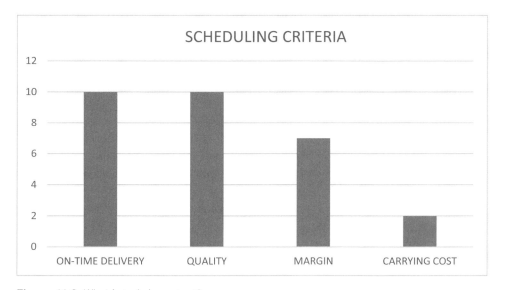

Figure 11.3 What is truly important?

quality, margin, carrying cost—I will call this the DQMC method. Quality is part of this decision tree to prohibit scheduling personnel from deviating from the primary process by changing workcenters, subcontracting, or substituting raw materials. While these practices may create a part to the print, the unnecessary introduction of variation will increase risk. These steps are sometimes necessary but should not be made or managed by scheduling personnel alone.

If a customer requires ten pieces per month for four consecutive months beginning in sixty days, is your lot size 10, 20, 30, or other? I will provide a technical answer: It depends! The answer depends on the variables we have been discussing: lead time, total setup time, number of operations, outside processes, minimum lot charges, and inspection frequency, whether customer specified AQL or internal requirements. Use of the AQL chart does not mean that you are willing to ship some parts not to print. It is a statistical document that determines the amount of measurement data the customer expects to be collected. The requirement and intent is always 100% quality. If the raw material is available and the pieces may be completed in a few operations in a cell, you will run ten at a time when each order is due to the customer—this is optimal and thus the goal. Conversely, if these components require ten manufacturing operations, two outside processes, and twenty hours of setup, then the order quantity is likely to be 40 or potentially split into an order of 10 and an order of 30 (see Figure 11.4). The carrying cost for 30 days is insignificant as compared to the long-term effects of missing a customer delivery, and insignificant compared to the effects on margins of the other decision parameters.

There are two other alternatives to discuss. First, if it is the initial time this part number will be manufactured, then it is possible that the lot sizes would be 11 and then thirty. It is wise to comprehend that one piece will be required for setup and debug of the machining processes. If you only run 40 total and have one piece of fall-out, then you will be one piece short. This may require new material to be obtained for a later production order of only one piece. The duplicate setup time incurred and outside-process lot charges required to manufacture this single component will certainly create a large negative margin for the 40-piece order. It is probable that the cost to produce this one piece will be equivalent to the total cost of 3–4 pieces manufactured in a larger quantity. This exorbitant cost of a single piece is the reality of complex machining that requires custom fixtures and dozens of unique tools to be set up for each operation.

The second lot-size alternative for complex components is to release four production orders of ten pieces simultaneously and obtain all the raw material at the same time. This provides the option of one, two, or all of the orders to be grouped as they move through the shop. If the schedule permits these orders to be machined back to back, then setups will be eliminated. After the third or fourth operation the schedule at a given workcenter may require the first order of ten to be

LOT SIZE PREDICAMENT

DELIVERY	DEMAND QUANTITY	LOT SIZE CHOICES	SELLING PRICE EACH	MATERIAL EACH	RUN COST EACH	OUTSIDE PROCESS EACH	OUTSIDE PROCESS MINIMUM LOT CHARGE	INSPECTION AQL COST TOTAL	TOTAL SETUP COSTS	TOTAL REVENUE	COST	PROFIT
OCTOBER	10	10	$ 500	50	$ 200	50	$ 1,000	$ 500	$ 1,200	$ 5,000	$ 5,700	$ (700.00)
NOVEMBER	10	20	$ 500	50	$ 200	50	$ 1,000	$ 1,000	$ 1,200	$ 10,000	$ 9,200	$ 800.00
DECEMBER	10	30	$ 500	50	$ 200	50	$ 1,000	$ 1,000	$ 1,200	$ 15,000	$ 12,200	$ 2,800.00
JANUARY	10	40	$ 500	50	$ 200	50	$ 1,000	$ 1,000	$ 1,200	$ 20,000	$ 15,200	$ 4,800.00

Figure 11.4

completed and moved forward, while the remaining orders will be machined a few days later.

It should be noted that producing some of these components before the ship date is not considered "make to stock", but rather "make to order". The key difference is that when you make to order you have a contract (the PO) from your customer to purchase the items. There is no risk. In the one-in-a-thousand chance that the order is cancelled, the customer will pay any cost already incurred. It is more expected that the customer will request early shipments. The early shipments may be provided at no extra charge, or there may be times when an expedite fee will be negotiated. The exercise to determine lot quantities is a customer satisfaction and profit maximization exercise. Figure 11.4 demonstrates that in nearly 100 % of the scenarios the fixed cost of setup, outside processing fees, transportation, paperwork, and inspection are higher than the variable cost of holding the inventory for a few weeks. For clarity purposes I want to emphasize the difference between long-term inventory with no order and short-term inventory with a contractual PO.

The predominance of machined components are actually small enough to be stored on a shelf or inside a cabinet. There are no spoilage, pilferage, transportation, or warehousing expenditures. In fact, handling is reduced by stocking one shipment instead of stocking three or four shipments. Again, always work to reduce setup and lead time, but I have demonstrated this scenario is common in today's manufacturing and will be more prevalent in the future. We need to understand that this is the type of work available to most Western manufacturing companies, and intelligent order sizes using the DQMC method is how we are going to be successful in this lower-volume and higher-complexity environment.

While it may be less prevalent, there are still some manufacturing organizations that release orders for stock with excessive lot quantities and face obsolete inventory predicaments while also consuming capacity. Subsequently, the normal reaction is a change in either management personnel or a philosophy that over-compensates with order quantities too small, which are not matched to the current Lean status of the company or simply not matched with the innate complexity of the part or the customer ordering pattern. The inevitable result is machine and human capacity consumed with setup and teardown, while support personnel are inundated with an exponential increase in transactional and physical handling of additional orders. On-time shipments are reduced, while costs increase and customers are irate. I compare the organizational phenomena of these extreme swings in order quantities to a pendulum on a cuckoo clock. The pendulum tends to swing from one side to the other side every few years, depending on the latest pain and who is in charge. The wise approach is to keep the pendulum closer to the middle, as it relates to your products and your current circumstances.

lean philosophy recommends "one-piece flow". This literally means to manufacture one piece and move it to the next operation. This forces work cells with operations in close physical proximity. The benefits are low inventories and lower scrap and rework. The scrap/rework advantage can be significant. The philosophy can be summarized as follows: Even if the company were to make the same number of mistakes with the one-piece-flow concept, there will be less bad parts, because the order quantity is one. Since the mistake is discovered faster as the part is moved to subsequent operations and becomes a completed component, the result of each mistake is a lower dollar impact, because less pieces should be effected.

The lower-lot size theory is based on the mathematical fact that an order of one piece creates the shortest total cycle time to deliver a part from the first operation through the last operation. The easiest example is a part with three operations that each requires exactly two minutes. If one part is introduced into an empty cell, it will proceed through the three operations in six minutes. If a lot quantity of two pieces is introduced into the cell, this lot will require four minutes at each operation and twelve minutes to be all the way completed. A lot size of five would require 30 minutes (see Figure 11.5). This logic assumes that each operation is immediately adjacent to the next and that there is no transportation time. It also assumes that there is one person at each operation and that there is no setup. The total time per piece is always six minutes, whether processed in large or small lot sizes.

LOT QUANTITY	OPERATION 1	OPERATION 2	OPERATION 3	TOTAL CYCLE TIME
1	2	2	2	6
2	4	4	4	12
5	10	10	10	30

ALL TIME IN MINUTES

Figure 11.5 One piece flow

How does this theoretical lean flow affect your decision on lot size? First, if you have this simple model where the same item runs repeatedly and you have one person at each station, then run the theoretical minimum of one. A one-piece flow is more common and feasible with assemblies and simple, small, fabricated components that can be manufactured in a cell. If your parts and/or equipment are not so small that you can hand the component to the next operation, then this method will not work. I would suggest running only the quantity on order from your customer.

With some innovation you can obtain the benefits of one-piece flow and the benefits of a larger lot size. With the proper equipment, workholding, and programming it is sometimes possible to machine the component completely in one operation.

The best approaches are yielded from combination machines such as the mill-turn machines or machining centers with turning capabilities. Five-axis machining centers can access five sides of the part, and the sixth side can be a separate setup on the side of the table so that one program completes all six sides.

Even with traditional vertical machining centers there is the capability to establish multiple fixtures on the table, so a single component can be rotated through several setups. If the part requires four setups, they can all be accomplished at one time, and a single program would machine all four setups. Each time the door is opened one complete part is removed.

Regardless of the type of machine tool, the more machining you can achieve with each cycle, the higher your total productivity. Innovative workholding and programming can allow batch production to minimize outside process charges and complex setups while simultaneously permitting a one-piece flow by this one-and-done approach.

There is a difference between writing about scheduling/inventory or consulting about scheduling/inventory and being responsible to customers, employees, and shareholders for executing the schedule. In the real world there are west coast dock strikes, earthquakes in Japan, supplier facilities that burn, crippling winter storms, hurricanes, unexpected bankruptcies, quality shutdowns, and many other unforeseeable predicaments. According to Standard and Davis in Running Today's Factory, "... academics rapidly discovered that it is easier (and safer) to solve problems that had no relevance to reality" (Standard and Davis, 47).

Even Toyota has been criticized by lean experts for allowing inventory turns to be reduced after 1990. It does not matter that Toyota doubled their international footprint, doubled sales, and accomplished their goal of becoming the world's number one automaker. With longer supply lines and a desire to avoid risk it was likely an intentional decision to have more supply in the pipeline. It seems that some Lean experts believe that companies are in business to compete on inventory turns.

CAPACITY MANAGEMENT

Advantages derived from your scheduling system will be mitigated or eliminated if not paired with a capacity management program. There are three steps to effectively establish a capacity tool that can transform your capacity understanding from a black hole to a weapon that delivers existing orders on time and wins new orders:

1. Determine hours of current capacity available at each workcenter. Hours may be machine hours, man hours, or both. Some facilities and processes are governed by the machine, and others are governed by the human. There is no right answer, but this needs to match the hours displayed on your routers and production orders. If your ERP and labor reporting can accept and measure both, that is

excellent, but you cannot do capacity in machine hours and production orders in man hours. There are methods to make one number measure both, but I will leave that for a future discussion.

Rarely is capacity a macro-problem, but frequently a micro-problem. It does not matter how many assemblers and machining center people are available when you are facing a bottleneck in precision grinding. Your capacity measurements need to be delineated to the necessary degree, if they are to be useful. Each workcenter should have available hours loaded into your ERP system or other software utilized for capacity. For reporting and graphical purposes it is practical to group some similar workcenters and then drill down as required.

A decision needs to be made to establish how many hours per week one employee will count in your available capacity projection. The variables to include are: overtime, utilization, efficiency, vacation, and any other time away from work. I recommend conservatism when calculating available capacity, as it is better to have a little more hours than projected rather than to be shorter than projected. An example would be to project no overtime, 90% utilization, and 90% efficiency—hence, 40 hrs. X .9 X .9 = 32.4 hours of actual work accomplished per employee per week at the given workcenter. There will be weeks with bereavement, vacations, equipment problems, FMLA, etc. Establish a reasonable calculation and move forward. You can tweak the formula in the future if required. As machinists are added or subtracted, your ERP should be updated.

2. Determine the number of hours by workcenter to complete existing demand. This is accomplished through the routing integrity building block of our Machining Pyramid. When the routing is constructed, a setup and runtime is determined for each operation. Whether the existing demand for a given part number is a quantity of 1, 10, or 100, your capacity-planning software will allocate the setup hours plus the run hours multiplied by the demand quantity. This calculation provides the total hours per workcenter for the demand. Based on the order due date for the demand, the scheduling system will determine the day and week necessary for each operation to be completed. We now have the hours per workcenter and the specific week the order will be at the workcenter. If we aggregate all demand, we will have a capacity report showing total hours of demand, by workcenter, for each week and for each month.

3. Now that we have hours available at each workcenter per week (step 1) and hours required by workcenter per week to meet existing demand (step 2), we can compare and analyze. I recommend a graph for easy visual display (see Figure 11.6). You will need to determine the time horizon to report and graph. This will vary by organization and perhaps by customer or product. I prefer to display weekly for a 12-week span and also to consolidate monthly for a 12-month view.

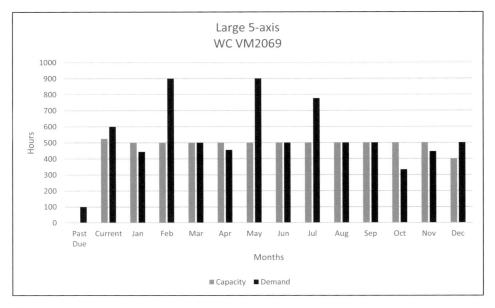

Figure 11.6

Now that you have a picture of your workcenter loading, what do you do next? I have chosen the term "capacity management" to suggest that companies should act on their knowledge of capacity issues to improve profit, delivery, and stress. The Machining Pyramid discusses many actions to practically manage capacity. Some of these are discussed in various degrees in other chapters.

a) If the demand hours are trending downward in certain workcenters, then work less overtime, transfer people to where they can be better utilized, and/or conduct cross-training.

b) When you have open capacity in specific workcenters, but not in others, push sales to fill the void(s) by selling part families that require the available workcenters. This may be accomplished through offering shorter delivery or more aggressive pricing.

c) When you have open capacity in specific workcenters and cannot accomplish steps 1 or 2 above, you should evaluate moving orders forward in the schedule. Since the vast majority of machine schedules are not level, this is sometimes referred to as cutting off a peak and moving forward into a valley. It is much better to move the peak forward than to move it backward.

d) Short-term overloads in specific workcenters can be overcome by first moving available cross-trained machinists into the area. Secondly, overtime can be added or increased.

e) Short-term overloads in specific workcenters can also be controlled by moving the work to other types of workcenters. Moving people to the work should always be the first choice, as moving the work requires re-processing, new programs, new routers, etc. This creates risk and will require support from engineering, programming, and quality. New customer projects may be jeopardized, as support personnel are distracted. Continuous improvement efforts may also be delayed. Overall, using valuable human assets to regularly reprocess work for different workcenters is not a winning formula. Finally, productivity should be lower when the work is moved to a different workcenter. The reason why the work was processed and routed to the original workcenters is that these were the most productive for the specific geometry.

f) Short-term overloads in specific workcenters may be alleviated by subcontracting to an outside company. This will depend on the complexity and the requirements of your certifications. Of course, strategic or proprietary processes and components should be kept internal.

g) Long-term overloads in specific workcenters need to be addressed through a project designed to prevent the capacity constraint or bottleneck. This may be accomplished through a productivity improvement project, a capital acquisition project, or by adding more shifts if feasible.

Larger and more sophisticated customers are seeking suppliers who understand and actively manage their capacity. They want to see and discuss your capacity-management program. Aerospace OEMs are pushing this issue heavily in their supplier development programs due to the long-term growth in their business. Regardless of the industry, your capacity management can be used as a sales tool. Customers who are sourcing critical projects, especially large projects consuming a high number of machining hours on rare workcenters or a specific workcenter, want to feel comfortable that their supplier has a complete grasp on the scheduling and the capacity necessary to be successful. When you can demonstrate where and when their project will hit on your capacity charts, the likelihood of receiving the order will be significantly increased. There have been many times that customers awarded me the order based on this information. Additionally, there were occasions that the customer intended to split a large order among multiple suppliers and reconsidered after viewing my capacity-management tools and understood that I had the available machine hours. These are important strategic wins, as it also keeps a competitor out of the project for future price and delivery pressure.

12 OPERATIONS MANAGEMENT

Operations management is the hub of the wheel in your organization. Even with modern equipment, bullet proof processes, and a suite of whiz bang software the company still needs to execute. "Execute" can be defined as utilizing resources to convert inputs to outputs within or surpassing cost, quality, and delivery targets. Execution also includes converting strategy into reality. Good strategy that is not implemented, not implemented correctly, or not implemented timely is meaningless. Organizations require people who can get things done. People who will remove barriers, go around barriers, or through the barriers. Which is more valuable and challenging—the executive who creates the business strategy or the operations management team that executes and implements the strategy? Since many more strategic plans fail due to implementation, my money is with the operations management team.

It is important to view the location of operations management in our Machining Pyramid—in the center. Below operations management is a solid foundation of resources: equipment, people, IT, leadership, quality, Lean, Six Sigma, routing integrity, and program validation. The operations management team is not only instrumental to developing and maintaining these foundational competencies, but they employ them 24/7. Adjacent to operations management in the Machining Pyramid is the supply chain and scheduling/capacity management. The operations management team works in parallel with the supply chain and the schedule to transform raw materials into finished goods.

Technology facilitates your competitor's ability to copy your products, your distribution, your software, and many other aspects of your business. World-class operations that cannot be copied are one of the few ways to sustain competitive advantage.

I consider the operations management team to not only include the production supervisors, factory management, and maintenance but also quality, materials management (supply chain), and manufacturing/process engineering. Materials management should include purchasing, production control, and shipping/receiving. Structuring these disciplines under one umbrella insures that all the necessary resources are aligned to fulfill customer requirements. Your team has built their Machining Pyramid, and it is the operations management folks who tactically execute all facets of the manufacturing process. This includes engineering, programming, scheduling, machining, assembly (or other internal operations), shipping, and the sourcing of materials, components, and outside processes.

It is also the operations management team that initiates continuous improvement to the Machining Pyramid and initiates developmental projects to fulfill customer needs prior to customers requesting the service.

Simply put, it is the operations management folks who have to run two races concurrently: They are running a sprint to take care of the customers this week, and they are running a marathon to develop people and processes the customers will need in the next 3–5 years.

I will examine some of the operations-management-team key responsibilities further:

PERSONNEL DEVELOPMENT

We discussed the importance of the workforce in the People building block of our Machining Pyramid foundation. It is the operations management group that is responsible for the day-to-day interface that enhances the character and motivation of your empowered teams.

The development process initiates during the hiring phase. Each organization has its own approach and philosophy to interviewing, selection, and development. I believe that each hire is an opportunity to either increase the talent and character level or to lower the talent and character level, whether the position is shipping, machining, fork lift driving, engineering, or an executive. You do not want to saddle your managers with a new person that creates quality concerns, because they are not conscientious, stray from their work area, surf the net, or display attendance issues—left unchecked, these behaviors can permeate your culture.

The goal is to hire people that cannot only be successful at the open position you are seeking to fill, but also to hire people that contain the potential and motivation to increase their value and be ready to step into future openings. All winning teams need good players on the bench, and you can never have too deep a bench.

Every organization will make a few bad hires, and every organization should be able to absorb or recover from some bad hiring decisions. However, over a period of time the organization that makes better hiring decisions will be the better organization. The most salient recommendations I can provide for the hiring process is to always utilize multiple people for the interview, insure that at least one person is highly knowledgeable of the skills required for the particular position, and that at least one interviewer has a high degree of emotional intelligence. We all have our strengths and weaknesses. Many people are just not perceptive at reading candidates to determine sincerity, ambiguity, confidence, high motivation, low motivation, or inconsistencies. It is said that Oliver Cromwell could read men like most people read books. Every organization needs an Oliver Cromwell.

If you need to invest months training engineers or programmers, you need confidence that the candidate not only possesses the technical skills but that he/she also is a cultural fit within the company. If you are going to train a machinist candidate for months or years prior to moving him/her to a nightshift, you really need to understand his/her personal situation to know he/she will not leave before or shortly after moving to nights. You need to discuss overtime requirements, pay, commute times, and benefits to understand whether the candidate is a fit or whether he/she is merely looking at your organization as a temporary paycheck until he/she can obtain a position that better suits his/her personal requirements. A good interviewer will ask the pertinent question, observe and listen, then ask the logical and probing follow-up questions. Ascertaining a clear picture of the candidate's skill set, intellectual capacity, motivations, stability, and suitability to your position is an invaluable asset. The person or persons with this aptitude need to be a key part of your leadership team.

You may learn more during a plant tour, testing, and/or reference checks. These each have pros and cons that I will leave to each organization to determine. Nothing will be more effective than multiple interviews and an interviewer with emotional intelligence.

I have seen many companies use one manager to perform the interviewing, especially for manufacturing positions. Frequently, the manager is inexperienced and ineffective at interviewing, and their poor hiring decisions have proven to be costly and damaging to morale. Why are bad hiring decisions unhealthy for morale and the culture? First, every company has a limited number of veterans who are capable and willing to train the rookies. When these veteran trainers see good employees leave due to pay, shifts, commuting time, job assignments, etc., they become frustrated. When these trainers are given people lacking motivation, work ethic, or the intellectual capacity to be successful, they become frustrated. Your trainers will become less enthused and less effective with each new trainee.

Compounding the problem is the afore-mentioned growth and baby boomer retirement, which requires most organizations to perform considerable training. The delta between hiring an inspired long-term fit and a short-term misfit is a greater impact to your morale and culture than meets the surface.

Operations management needs to excel at developing people. Developing is more than training on new skills. I favor what I have termed "constant stream development" (CSD). Development may be planned and formal, but it also needs to be ongoing and informal. Managers should insure that employees are exposed to new challenges, technologies, suppliers, and/or customers. This holds true whether the person is assigned to the boardroom, office, or the shop. Everyone is energized when asked to assist in a Kaizen team or other problem solving. Every employee can benefit by obtaining a wider perspective of the business. This type of development is a series of brief assignments that include representing the individual's department at internal meetings, supplier visits, presenting to customers, selecting new equipment, evaluating software, etc. The individual will continue with his/her current job function, while learning more about the business and providing decision makers more opportunity to evaluate his/her long-term potential.

The next phase of development is cross-training. We all have lost key employees very unexpectedly, and at times we have lost multiple employees in a short period of time within the same job function. As leaders, we need to insure that all technical and administrative functions have trained personnel fully capable of replacing any employee who quits, gets fired, wins the lottery or moves across the country to marry their internet sweetheart. Cross-training also resolves issues related to vacations and high-growth or capacity surges affecting specific disciplines. Cross-training your people is another opportunity for management to learn more about the individual as well as to provide opportunities for growth.

Many managers never cross-train until the situation is an emergency. Once the emergency arises, the training is ineffectual and overly stressful. Managers provide constant excuses for lack of cross-training such as cost, lack of time, or "they will lose the knowledge, if we move them out of the position". Good managers and leaders incorporate cross-training and knowledge retention as part of their core values and as part of the organizational culture. Even busy organizations occasionally experience slow times in engineering, machining, assembly, or maintenance. When slow times occur, it is much better to pull someone to cross-train than let the entire group coast. This maintains good work habits. An easy example is a work cell of five employees producing 200 units per day. If the volume diminishes to 160 units per day, the correct move is to send one person somewhere to cross-train, instead of allowing the entire cell to work at a slower pace. This example is applicable anywhere in the organization. If a manager cannot cross-train during slack time, he better force or create a situation conducive to cross-training.

So we have three types of routine development: the brief extra-curricular assignments, cross-training, and a lateral or promotional move to a complimentary position that will provide new challenges and growth. This type of development can be accomplished by large and small organizations. This type of development occurs within the normal flow of operations and is not disruptive. This constant stream of development helps establish a learning culture, keeps employees involved, and actually simplifies day-to-day operations, because managers have a broader and deeper pool of talent.

Managers should be evaluated and compensated, in large part, according to their ability to attract, develop, and maintain talent. Managers should always strive to identify their people who possess the potential to contribute either at a higher level or at an area of the organization that is experiencing a talent shortage. Practicing CSD either reinforces the belief in an individual's potential or demonstrates that the candidate is not ready. Good leaders are adroit at picking good candidates, at communicating, and at motivating them not only throughout the development assignments but throughout their career.

I am not recommending moving high-potential candidates between disparate functions such as accounting, marketing, and operations management. I addressed this during the Leadership chapter as a failed strategy that hinders development while creating morale and execution complications. Since the original pool of candidates does have significant potential, there are, of course, bound to be some success stories associated with this method. In aggregate, there are more negative outcomes and poor long-term ramifications that outweigh any success. When I speak of development assignments, I am recommending that proven performers, based on merit, be given a complimentary assignment that allows him/her to use his/her existing skill set while continuing to be challenged and grow. This employee can contribute in the new role instead of being carried by his/her new peers.

SCHEDULE ATTAINMENT

The operations management team directs and controls the resources to meet internal and external customer deliveries for products and services. The necessary resources include the supply base, scheduling/capacity management, equipment, and people. Since there is a separate Machining Pyramid chapter in which each of these topics is discussed in depth, I want to focus on the operations management team's role in utilizing these resources to successfully achieve schedule attainment without unnecessary waste such as excessive overtime, outsourcing, and expediting. Please note that I did not include inventory in the previous sentence. Too much inventory or inventory that becomes obsolete is, of course, a waste. Some inventory may indeed be a strategic decision for immediate response to a customer or to buffer a supply chain. Remember, not every business model is predictable like an assembly line, and not every supply chain member is beholden to the small and medium-size machining companies. In fact, most are not predictable, and 99% are less predictable than the assembly line at an automotive assembly facility. The vast majority of organizations do not have suppliers with a warehouse full of parts down the street. Most manufacturers in the developed economies have some level of sporadic demand. Many, such as those supplying replacement components for aerospace or defense have extreme swings in demand.

Whatever scheduling methodology you are employing, the most important step is that everyone interfaces with the system 100% as designed. I need to emphasize the words "everyone" and "100%" in the previous sentence. Every scheduling system will break down if not administered or managed properly. If a machining organization is processing 800 orders concurrently with an average of eight manufacturing sequences per order, there are 6400 discrete operations to manage at any given time. This does not include ordering the raw material, ordering purchased components, receiving transactions, shipping, and any internal material movements. There will always be some amount of schedule disruptions due to demand spikes, supplier shortages, manpower availability, bad weather, dock strikes, and many other common and uncommon circumstances. Your operations management team not only works to minimize these events, but they adjust to recover from these events. You need sharp, aggressive, innovative people who consistently act in the best interest of their own company and the best interest of the customer to overcome schedule problems without introducing a quality risk or compliance failure.

There are two types of scheduling failures: predictable and unpredictable. The unpredictable have already been partially discussed, i.e., bad weather, equipment failure, supplier surprises, scrap, etc. A process with a historical near 100%-yield that suddenly generates excessive scrap is also unpredictable. While unpredictable schedule complications are difficult to anticipate, the predictable scheduling complications should be anticipated. Predictable scheduling issues are primarily generated through product mix, timing, and volume. A manufacturing company may be fortunate to have level revenue and labor hours for all twelve months of the upcoming year. However, while there may not be any macro-capacity concerns, there may be a number of micro-capacity concerns. A typical organization may have 75 workcenters that comprise the total capacity of the facility. If one of these workcenters is a precision grinding operation that has 500 hours of capacity available each month but has 900 hours demand in months two and five, then you have a micro-capacity problem during these two months. This is predictable. Good companies have good capacity planning. Good capacity planning requires good execution in many disciplines, i.e., the Machining Pyramid. Great companies respond to their capacity planning and minimize the effects of the inevitable micro-bottlenecks generated by product mix, timing, and volume.

If the 900 hours of demand in the precision grinding workcenter consists of thirty individual orders, then not only will many of these orders be late to the customers, but the bottleneck will last well into the following month(s), creating dozens more late orders. This occurs, because the orders scheduled to be completed in month three cannot be completed on time due to the carryover of orders from month two (see Figure 12.1). If some of these late components are utilized in assemblies, the magnitude of the micro-bottleneck is magnified.

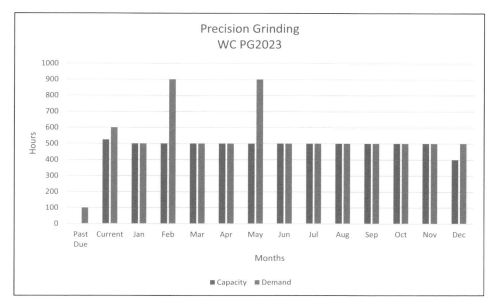

Figure 12.1

Great companies successfully maneuver through potential micro-bottleneck land mines, because they had the vision and leadership to develop alternatives. It is not easy to outsource precision operations, especially considering that many industries (aerospace, medical) have established frozen processes and require special certification from any machining sources.

Generals win battles due to their decisions and actions prior to the battle being fought. Ideally, your capacity planning should allow you to perform capacity management and avoid the predictable bottleneck. Unfortunately, due to a combination of predictable and unpredictable issues you will inevitably find yourself in the midst of a bottleneck.

There are a variety of approaches to staffing nightshift and weekends to increase throughput and provide flexibility for schedule attainment. Additional shifts require considerable training of not only machinists but the possible support people (programmers, inspectors, maintenance, management), depending on the size and complexity of your operations. These additional shifts also require much more management attention. Nightshift or weekend shift personnel tend to be less permanent due to a desire to be on dayshift, or they may simply be more transient by nature. Multiple shifts require exponentially more communication, not just for routine HR issues but daily micro issues for each workcenter, related to work instructions, quality concerns, maintenance problems, etc. Effectively managing multiple shifts requires superior leadership and the benefits derived from your Machining

Pyramid such as robust processes, standardization, routing integrity, in-process inspection, etc. The worst decisions about multiple shifts would be a decision to eliminate, reduce, or not instigate.

Many small organizations shy away from the challenge of nightshifts and hence will remain small or be eliminated at a future economic downturn. The quality and productivity of a nightshift is too difficult, because the small company does not practice the Machining Pyramid.

Equipment and automation are easier to justify, if you amortize over more hours per week and hence more hours per year. Let us compare two organizations each with $900,000 in capital to spend for the upcoming year. Both companies utilize the same type of equipment consisting of fifty lathes and mills of various types. Company A operates one shift and must purchase two more CNC mills and one more CNC lathe identical to existing equipment in order to meet current demand. Company B operates two shifts with overtime and operates all primary equipment 110 hours per week. Company B has no bottlenecks or constraints to meeting current demands. Company B is able to purchase different types of equipment to execute their strategic plan and fulfill expanding customer needs for higher precision and complex components. Company B purchases two wire EDM machines and one precision ID/OD grinder. The next year Company B purchases an automated cell for its largest part family. There is no question that Company B will be more diverse, profitable, and customer-focused. The moral of this story is that buying new equipment to increase your current capacity for existing underutilized equipment is a financial and strategic blunder. If you are an owner, president, or vice president being asked to authorize more of the same type of equipment that your data and eyes tell you is not reasonably utilized, you have a problem.

If you are achieving your schedule easily, then you probably have too much inventory or too much capacity. Unless you have a very simple product, steady demand, and minimum product mix, your schedule will be challenging. Manufacturers attempt to match capacity to demand, which often leaves little room to meet demand spikes and overcome supply disruptions. A certain amount of inventory is often a strategic intent for many business models and will assist in leveling wide demand and capacity deltas. Inevitably, all companies face bottlenecks in their supply chain both internal and external. My experience highly meshes with Goldratt's "Theory of Constraints", which suggests that you should micro-manage your bottlenecks. The Machining Pyramid provides a full set of tools that permits the operations management team the ability to identify bottlenecks before they occur and eliminate their potential damage. Likewise, once a bottleneck occurs, the Machining Pyramid provides real-time identification and the agile capabilities to move the people to the work or move the work to the people.

Some lean practitioners advocate ignoring bottlenecks, as they tend to "wander." I would argue that the "wandering" of a bottleneck is positive, as it reflects effective action has overcome the primary bottleneck. Micro-managing the bottleneck also includes improving productivity/throughput of the restricted process to prevent the same root cause of the bottleneck from occurring in the future.

PRODUCTIVITY

Productivity cures many ills. High productivity generally leads to solid profit. High productivity simplifies schedule attainment. And high productivity indicates minimum scrap/rework, as the latter prevents the former. Every operations management team is charged with measuring and improving productivity. Most people really cannot define productivity! In broad economic terms productivity is output/input. In a manufacturing setting the most meaningful measurement is: PRODUCTIVITY = UTILIZATION X EFFICIENCY. Utilization is a percentage indicating how much of the machinist's day was spent on the intended direct labor activity. If the person spent ten hours operating his machine tool and worked ten hours that day, his utilization would be 10/10 or 100%. If the next day the person spent nine hours operating his machine tool and one hour indirect time in a safety meeting, his utilization would be 9/10 or 90%. Utilization of equipment is also frequently measured in the same manner, in regards to labor, productivity utilization refers to the individual.

Efficiency is the amount of work completed, divided by the amount of hours spent directly to accomplish the work. The amount of hours spent to accomplish the work does not include any indirect time that is captured in the utilization calculation. If the measured or historical standard to assemble/machine/weld one unit is one hour, then the employee who completes one unit per hour is 100% efficient. If the employee spends ten hours to assemble/machine/weld the units and only completes eight units, his efficiency is 8/10 or 80%.

In our example of utilization the employee works ten hours, but attends a safety meeting for one hour. He only has nine hours of direct labor time to assemble/machine/weld. If he completes seven units, his efficiency is 7/9 or 78%. Therefore, productivity = .9 (utilization) X .78 (efficiency) or 70%.

As you can see from this example, it is important for utilization to remain high. If employees intended to perform direct labor activities are instead performing indirect activities such as maintenance, cleaning, searching for fixtures/tools, etc., you will have low productivity. Training is generally the only legitimate indirect activity that will become an investment.

You may ask why you should bother to measure both utilization and efficiency to obtain a productivity metric. Would it not be easier just to measure how many units were produced per total hours worked? The answer is that we need to create metrics with enough granularity, so that we understand root causes to perform short- and long-term actions for continuous improvement. Even though we are measuring the utilization of the machinist, the metric reflects management effectiveness. Equipment failures, misplaced fixtures/tooling, CNC program malfunctions, etc. are predominantly not the fault of the individual machinist. The machinist will record any portion of his day spent on these activities as indirect labor, which will permit metrics for each type of indirect labor per department, shift, supervisor, or individual. Different indirect codes can be created so that the various types of indirect activity (training, maintenance, cleaning, tooling, etc.) can be measured. This can be summarized and trended for any date range.

Since each direct labor person can record their non-value-added activities as indirect labor, the remaining hours are clearly spent performing direct labor activities. The efficiency metric will measure how much the individual completes during these hours. An individual's efficiency is therefore clearly within his own control. He cannot blame malfunctioning programs or missing tooling, as these activities are part of indirect labor and measured separately by utilization. This permits those with higher efficiencies to receive an increase or other types of rewards and reinforcement. Individuals will learn to discuss problems with efficiencies immediately with their supervisor to correct problems in real time. This reward and feedback system not only promotes continuous improvement but increases communication about problems with the manufacturing operation or the amount of time required to complete the operation. The time estimates to complete each manufacturing operation for each part number will become more accurate, and there will be less variance between the estimate or standard and the actual time required to complete the work. Low variance between the estimate/standard leads to more accurate scheduling and capacity planning. Accurate and historical data will lead to more accurate quotes, which yield fewer losers and more winners. Companies without good data end up with the losers.

Individuals who encounter a problem that requires them to stop working on their direct labor assignment and start on indirect labor are instructed to notify their supervisor whether the problem cannot be resolved in ten minutes. This insures that the correct resources are immediately applied to resolve a problem with a sense of urgency or that the supervisor can redeploy the person until the problem has been resolved.

This data can be easily collected electronically and summarized as shown in Figure 12.2. It is important to reinforce that profit is a measure of the past, but productivity, utilization, and efficiency can be measured immediately. This real-time information facilitates continuous improvement and drives scheduling accuracy, costing, and capacity management.

Figure 12.2

Report - Labor Efficiency per Employee per Supervisor
ACME INC.

Date/Time   2/14/14 [01:00] to 2/14/14 [23:59]
Department/Supervisor: 1234 Apple, John   Supervisor: John Apple

Employee Name	Item #	Prod. Order	WC Desc.	Sequence	Work Ctr	D/I	S/R	Routed Hrs.	Qty Com	Actual Hr	Earned Hrs	Efficiency
624 Kevin Stop												
	72078353	702126	CNC Lathe	130	L12	D	S	1.50	1.00	2.20	1.50	68.18%
	163-277	702126	CNC Lathe	111	L12	D	R	0.40	14.00	5.10	5.60	109.80%
	163-287	702242	CNC Lathe	100	l12	D	S	1.25	1.00	1.25	1.25	100.00%
	FIRST ART	FIRST ART	FIRST ARTICLE INSP.			I		0.00	0.00%	0.45	0.00	0.00%

Totals for Kevin Stop:

Total Hrs	Indirect Hr	Util%	Direct Hr	Earned	Efficiency
9.00	0.45	95.00%	9.00	8.35	92.78%

Employee Name	Item #	Prod. Order	WC Desc.	Sequence	Work Ctr	D/I	S/R	Routed Hrs.	Qty Com	Actual Hr	Earned Hrs	Efficiency
545 Dan Everest												
	703136	1497696	OD Grind	120	262	D	R	0.45	10.00	4.22	4.50	106.64%
	70247	1146876	Surface Grind	110	A51	D	R	0.17	8.00	1.57	1.36	86.62%
	70248	7529376	ID Grind	80	A51	D	R	0.35	9.00	2.63	3.15	119.77%
	MEETING	MEETING	MEETING	10		I		0.00	0.00%	1.00	0.00	0.00%

Totals For Dan Everest:

Total Hrs	Indirect Hr	Util%	Direct Hr	Earned	Efficiency
8.50	1.00	88.24%	8.42	9.01	107.01%

Indirect Labor Hours:

	Total Hrs
ADMINISTRATIVE	9.20
CLEANUP	10.18
FIRST ARTICLE INSPECTION	3.10
MEETING	10.05
TRAINING	1.38
Total Hrs	405.67

SUMMARY FOR SUPERVISOR Apple, John

Indirect Hr	Util%	Direct Hr	Earned	Efficiency
33.91	91.64%	351.38	345.22	98.25%

SUMMARY FOR Department 1234 Direct Labor 4

Total Hrs	Indirect Hr	Util%	Direct Hr	Earned	Efficiency
405.67	54.28	86.62%	351.38	595.97	169.61%

Legend:
Earned = (Quantity X Run Time) + Setup Time
Actual = Run time + Setup time

Efficiency = Earned / Actual
Utilization = Tot Hrs - Indirect/Tot Hrs
D/I = Direct, Indirect

S/R = Setup, Run
Oper = Work Order Routing Oper.
WCT = Work Center

The first step to improvement is creating an accurate metric. Productivity in the aggregate is output divided by input for your entire workforce, as mentioned earlier. However, for machining as in most manufacturing that does not utilize an assembly line the metric begins by measuring each individual employee. A machining facility with 100 machinists may very well have no two people performing the same activity. Each machinist may be operating a different type of machine or machining a different component or a different operation on the same component. Many of these machinists will be assigned a different component or a different operation the next day. Understanding how much work should be completed each day and how much work was actually completed each day requires careful planning.

As shown in our example, the output of an individual is the number of pieces produced times the standard per piece, and the input is of course the number of direct labor hours expended to produce the output. The aggregate workforce is all direct labor hours on all shifts including all direct labor overtime hours worked. Measurement of utilization, efficiency, and productivity equally for all shifts and all employees establishes accountability for everyone. This includes machinists, inspectors, CNC programmers, and operations management. Strengths, weaknesses, and trends are all on display in these metrics. Given that the Machining Pyramid provides accurate routings, accurate time estimates, bar coded production orders, closed loop ERP tracking of quantities at each operation, and comprehensive labor reporting, there is no guessing, wondering, or hiding any individual or group productivity.

There are variations to how the data may be collected for cells and high-volume manufacturing. There are also third-party machine monitoring software packages that collect extensive process data. What is a necessity is that your data is accurate, timely, stratified, and in a format to display individual, group, and shift trends. This permits regular action to correct and improve micro- and macro-issues. There are many drivers for productivity in a machining facility. We have discussed nearly all of them: equipment, robust processes, training, quality, etc. The last piece of the puzzle is simply the human effort and human time management by both the machinists and the operations management team. You may have the best equipment, the best metrology, and a robust process, but if the human equation is lacking, your performance will be lacking. The people still need to execute each day to obtain the potential of full productivity. The operations management team must insure each hour, shift, day, and month that the entire workforce is engaged, motivated, managing their own time, and that the spindles are making chips.

Besides measuring people, utilization can also be employed to measure what percent of the day equipment is running. While we want people to be utilized close to 100% of the day, the equipment utilization calculation requires more discussion. When equipment is scheduled to operate more than 80%, it is more probable for bottlenecks, WIP, and late deliveries to increase. It is advantageous for high equipment utilization in the 80–90% range, but there need to be real-time metrics to evaluate conditions leading to late deliveries and overtime. I discuss this in detail in the Scheduling and Capacity Management chapter. Some experts indicate that profits are lowered when equipment utilization is too high and advocate for lower utilization. The fallacy of this concept is that low utilization does not produce revenue. High utilization, effectively managed, will produce higher revenue and more overhead absorption. When capital equipment such as machine tools cost six and seven figures, equipment utilization is more critical than in lower-capital industries. Proper planning, management, and leadership can yield high equipment utilization and high profits.

Just as utilization can be employed to measure either the human or the machine, so can productivity be employed to measure the human or the overall process. So far we have discussed productivity of the person. This measurement implies how productive the person was with the tools and process provided. What if one person was provided tools that dramatically increased the potential for higher output? For example, a construction worker creating a temporary drainage ditch with a shovel worked very hard and created a ditch 2' deep and 75' long in 20 hours. Before the next job, the construction manager purchased a bulldozer and created the same ditch in 1.5 hours. The productivity was much higher with the introduction of the automation provided by the capital investment of the bulldozer.

Machining companies have varying levels of capital invested into their equipment and varying levels of successful automation achieved with these investments. Chapter 3 introduced a concept of a productivity multiplier. The total productivity of a process is the combination of human skill, human effort, and the level of capital equipment automation. This is represented by the equation KI X ME X C = total productivity. Skill is represented by KI (knowledge & innovation), effort by ME (motivation & effort), and capital by C. The formula acknowledges that capital investment by itself is not the answer. Likewise, a skilled employee must have more than his innate aptitude and acquired knowledge to be valuable. He must also be motivated to perform and provide the effort to actually perform. Without the requisite KI and ME the C will vastly under-produce both quantity and quality.

Figure 12.3 demonstrates levels of productive capacity for each of our three variables. KI begins with an operator with minimal ability to load/unload and progresses to an operator/machinist changing tools and utilizing measurement instruments through a machinist's debugging programs and performing CI and process innovation. ME starts with a person watching the process and moves upward with the use of gaging and deburr tools during the cycle time to preparing for the next setup, running multiple machines, and finally time management and effort to run multiple machines and also measure components and prepare for next setups during the cycle times. C initiates with the simplest and basic manual machines and escalates with CNC machines to advanced CNC equipment, combination machines such as mill/turn, and finally equipment with auxiliary devices such as tool setters, automation, and in-process metrology.

	PRODUCTIVITY MULTIPLIER		
SCORE	MACHINIST		CAPITAL
	KNOWLEDGE/INNOVATION	MOTIVATION/EFFORT	
1	LOAD/UNLOAD ONLY	OBSERVE	MANUAL MACHINE TOOLS
1	CHANGE TOOLING	MONITOR	ENTRY LEVEL CNC
2	CHANGE TOOLING & OFFSETS	ATTRIBUTE GAGING	PRODUCTION CNC
2	ATTRIBUTE GAGING	INTERNAL DEBUR	FOUR AXIS
3	METROLOGY INSTRUMENTS	INTERNAL METROLOGY	HIGH PRECISION
3	SETUPS	INTERNAL SETUP	FIVE AXIS
4	DEBUG PROGRAM	MULTIPLE MACHINES	COMBINATION EQUIPMENT
4	PROBLEM SOLVE	MULTIPLE MACHINES INTERNAL METROLOGY	AUTOMATION
5	CONTINUOUS IMPROVEMENT	MULTIPLE MACHINES INTERNAL SETUPS	AUTOMATION W/AUXILIARY PROCESSES
5	INNOVATIVE PROCESS DEVELOPMENT	MULTIPLE MACHINES INTERNAL SETUPS/METROLOGY	AUTOMATION W/AUXILIARY & METROLOGY

PRODUCTIVITY = KI * ME * C

Figure 12.3

Choose your current score for your level of productive capacity in each category. The score may range from 1–125. This seems to be a high ratio, but it is evident that a bulldozer and truck can move an exponential amount of dirt X miles further than a human with a shovel and wheelbarrow, or that a bicycle, car, and plane can transport a human exponentially further than a pair of legs in one day. A motivated and talented machinist with two modern machine tools employing automation and auxiliary devices will be exponentially more productive in comparison to a low-tech manual machine with an entry level operator.

QUALITY MANAGEMENT

Just as the Machining Pyramid provides the tools for productivity, so does the Machining Pyramid provide the tools to produce quality products and services. Somebody needs to steer the quality ship, and that responsibility lies with the operations management team. The quality director and his people are part of the operations management team, but everyone in the company is responsible for quality. It is an important concept that engineers, managers, programmers, and machinists are just as responsible for quality as the quality department. Every decision made throughout the organization has some impact on quality. If every decision made incorporates this understanding, then achieving a quality culture and building quality into the product becomes much easier.

When a machinist notifies his supervisor that he is encountering difficulties with certain dimensions measuring just barely within the print tolerance, the supervisor's response is a barometer for management commitment to quality. Many supervisors will respond, "do the best you can" or "that dimension is not critical, so keep running". The correct response is, "let's see what we can do to get that into the middle of the tolerance". A technically oriented supervisor may be able to resolve the issue himself, or he may need support. The effort of the support team should be immediate. Directionally, this activity needs to be more like a NASCAR pit crew than a Six Sigma project. Even with a delay to improve the quality of a given dimension the order will probably be completed faster than if the problem had not been addressed, because the machinist does not have to fight the borderline quality implications. Without these improvements it is very likely that scrap or rework would have been created. Without the improvements it is very likely that the machining operation would experience the complications every time it is set up.

The decisions and behavior of the operations management team sets the tone for the entire organization on the commitment to quality. When all employees see the effort and dollars being spent to improve gauging, tooling, programming, etc., then they will also make the extra effort. This leads to more innovative and conscientious employees. This fosters continuous improvement.

Just as productivity requires accurate and actionable metrics, so does quality. Quality requires micro-metrics about individual parts, dimensions, machines, and employees. Quality also requires macro-metrics of scrap, rework, PPM, and cost of quality. Micro-metrics include capability measurements (CPk, Cp, Cr), gage R&Rs, histograms, SPC, and NCR data.

CONTINUOUS IMPROVEMENT

Your competition is improving! You must help your organization improve at a faster rate than your competition. You possibly do not know all your competitors and you can expect to have new competitors in the future. These are simple but unmistakable facts. Continuous improvement is not an option—it is survival! Improvements can come in big chunks or very small pieces. Improvements need to occur in all areas of your organization from sales and engineering to purchasing and shipping. Continuous improvement is not just for manufacturing. Continuous improvement needs to be part of your culture that emanates from an agile learning environment.

Actions that create CI take many forms: corrective actions, preventive actions, and projects. While corrective actions are initiated in response to a non-conformance, the result is still an improvement. What is important is the mechanism and discipline to capture the non-conformance, to assign the responsibility to the proper department or individual, to implement the correction, and to apply the lesson learned to similar parts or processes. Many companies do not define a non-conformance and do not consistently record them when they occur. Non-conformances will happen, but the best companies turn the failure into an opportunity.

Preventive actions generally occur when a risk has been identified, but failures or a non-conformance has not materialized. The risk may be in the form of borderline quality or poor productivity. These risks may be identified through labor reporting, in-process inspection, SPC, tool wear data, machinist recognition of fixture/gage wear, or aesthetics. Again, the best organizations utilize Machining Pyramid tools that identify and capture risks in real time or near real time. Timeliness is an important concept and is another trait of outstanding organizations. Timely data and timely identification of risk are actionable. Stale data are not normally actionable. Elimination of risk through timely preventive action is a key indicator that your operations management team is functioning at a high level.

Projects or planned actions are designed to increase capacity, improve quality, replace obsolete equipment, add new processes or services, etc. These actions may be in support of business or operational strategic plans. The planned actions may be incrementally expensed budget purchases or capital projects. These projects are macro in scope and create CI across the organization or within a department, cell, or part family.

CI must also occur on the micro-scale—one part number and one operation at a time. Figure 12.4 demonstrates the cycle that occurs every time a part number is produced and the methodology to collect data, implement changes, verify improvement, update costing/pricing, collect data, implement more changes, and repeatedly drive for improved quality, costing, and lead times. This relentless approach to seize every opportunity for CI becomes an ingrained culture that separates the best machining companies from every other organization. Each time a production order for a given part number is manufactured, the cycle of CI demonstrated in Figure 12.4 should be performed, and hence this can become a cycle of near perfection. An organization that properly performs this cycle should not lose a product, family, or assembly to the competition. Machining companies need to review Figure 12.4 closely and ask themselves if they have the tools along with the people in position, with the time and motivation to perform this cycle of CI.

Regardless of the size or type of action, it is the operations management team that encourages, supports, and rewards CI effectiveness. It is the operations management team that will remove roadblocks, solve problems, and apply resources to push and finalize improvements. An operations management team that lives and breathes CI is irreplaceable. When all employees are refining their own actions

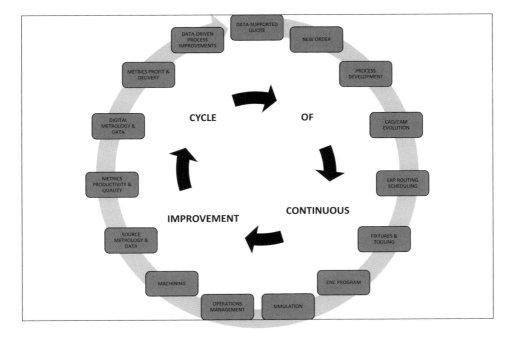

Figure 12.4 Part number specific – data driven – continuous improvement

and time management, the organization moves with a sense of urgency, is more effective, and creates processes (both shop and office) that are much more robust.

CULTURE

Can you describe the current culture at your company? What culture would you prefer? Is there a gap? Culture governs your organizational productivity, quality, attitude, teamwork, and customer responsiveness. The operations management team is most responsible for your culture.

Below are the factors that shape, mold, and reinforce organizational culture:

- Hiring: We have already discussed that each new hire is an opportunity to add to the long-term skill level of the facility. It is also an opportunity to reinforce your desired culture or introduce a person that subtracts from your desired culture. Each person's effort, skill, and performance can either be at a high mark for others to strive to achieve or a low mark for others to emulate.

- Personnel decisions: Regardless of the circumstances or environment, when positive behavior and performance are rewarded, the aggregate workforce will produce more of the behavior. When poor behavior and performance are eliminated or reduced, you will also produce less of the behavior.

- Termination: Separating a person from his employment is always a difficult task and not taken lightly. Whether the root cause is performance or lack of work, the termination should be performed thoughtfully and professionally. Still, the termination establishes clear lines of demarcation on acceptable performance and behavior. Most employees are energized by knowing that their overall performance and potential is recognized and that they work for an organization that rewards contributors and sets reasonable standards that will yield long-term organizational success. "A-players" attract "A-players", and "D-players" attract "D-players". A-players expect managers and leaders to remove or improve D-players. If the D-player is not eventually removed or improved, the problem is no longer the D-player. The problem is now his manager for failure to take appropriate action.

- Behavior: You cannot change the way people think, but you can change behavior through rewards, recognition, and reinforcement. If you change behavior positively over a period of time, you begin to change the culture.

- Actions: Employees see written management decisions and hear management decisions, but decisions are simply intentions. Actions are real and create a pattern.

- Results: Good intentions and working many hours are not a substitute for success and for results!

- Change: Do not implement new programs unless you have enough management support—above and below on the organizational chart—to follow through to success. Never, never, never. When key personnel are not sold on new projects, they tend to not encourage or enforce to the degree necessary to overcome the inertia to change. Whether the situation requires more communication, more education, more time, or just a bigger hammer, you better find the answer or not proceed. If you let programs fade away after implementation, you are conditioning the entire organization to ignore your future initiatives. You will make every future endeavor exponentially more difficult.

- Quality: We have already discussed that operations management is the captain of the quality ship in the manufacturing organization. I want to reinforce the magnitude that the daily actions and decisions of your operations management team stamps upon the culture of your company. What is the response from managers and leaders when quality problems arise that are not black and white? Many issues are indeed grey. Does your leadership say "ship", "keep running", or "the customer won't notice"? The culture and legend at a Toyota assembly plant is that anyone at any time can stop the line. In a machining facility there are machine tools to stop instead of the assembly line, but your people need to have the belief that they can stop any machine at any time. In fact, all your people need to possess the desire and caring to stop any machine at any time!

The quality problem may still be within specification and will not generate a return or quality escape. The problem may be aesthetic—that is acceptable but not ideal. The problem may be bad blends or large burrs that will create extra handwork downstream. Regardless of the type of potential defect, when leadership responds with a combination of technical resources (engineering, maintenance, programmers, etc.) and necessary expenditures (gages, equipment, fixtures, tooling) to eradicate the problem, a clear cultural reinforcement has been delivered. In the first do-nothing scenario the cultural message is, "we will wait to fire-fight until there is a clear problem, but until then don't be concerned". Of course, there is a strong likelihood that eventually the grey area slips into intermittent failure unnoticed sometime during the night, resulting in an expensive and/or embarrassing quality escape.

In the second pro-active scenario, leadership delivers a clear message that the organizational quality reputation is valuable and must be protected. The message is that a borderline quality problem is an opportunity to improve and an opportunity to create a more robust process. When the employee is complimented for recognizing the initial problem and asked for his input, you are instilling a culture of ownership and pride. Conversely, when the employee is told to "keep running", you are instilling a behavior to not report future concerns, and perhaps some of these concerns will turn out to be quite serious.

In the proactive scenario you are reinforcing employees to be aware and astute to not only their processes, but aware and astute to their total environment. In the do-nothing scenario you are numbing the senses. When I am given responsibility to lead facilities in turnaround situations I frequently hear, "I didn't say anything, because the company never did anything about it in the past". I insure that my words and actions immediately begin to reverse these types of statements. It does not need to be a turnaround situation to establish this mindset. A new manager needs to establish this culture from the beginning. This is a major step to the elimination of fire-fighting.

Over the long term the employee quality culture will mirror the management effort, management involvement, and management expectation of quality. The quality culture is strictly a leadership issue. Moreover, it is the easiest branch of the culture tree to correct. Good quality requires less effort and less stress from people. Robust processes require less work, less inspection, and less rework. Productivity, attendance, and skill level all require a longer journey.

WALK THE FLOOR/TALK THE FLOOR

It is important for leaders to be seen, but more important for leaders to see! Technical or non-technical support personnel, management, and executives should all walk the company and the shop. It is ok not to have a technical background. It does not require a technical background to recognize an improvement or retardation in 5S. It does not require a technical background to notice whether a machine tool is running or idle. Bill Parcells, the famous football coach, once said, "you are what your record says you are". The same can be said about your walk through the shop. A machine running is a "W", and a machine not running is an "L". Over a period of time, if you have too many "Ls", you are what your record says you are and you need to ask some questions. It is helpful for the boss to associate activity within his organization with the trends in the metrics. People's behavior does change when the boss is looking. It is advantageous for staff and other indirect personnel to view new projects, new equipment, and general changes. When you fly between two cities you do not notice much about the detail of the landscape. You are too high, too fast, and most probably above the clouds. When you drive between the same two cities you see much more. If you had the time to walk between the two cities you would be immersed in the landscape, architecture, climate, people, and many unpredictable details.

The only way to view and absorb more than what is absorbed during a routine walk is to walk and talk. Some leaders and support personnel may feel out of their comfort zone on the shop floor. You should not feel that you need to be strong technically to casually interface with machinists, inspectors, and assemblers. Even the company's best technical personnel are not fluid in all processes and technologies. In those rare cases when someone asks a technical question, a smile and a glib "hell I don't know that's what I pay you guys for" will suffice.

The more you move throughout the company, the more comfortable you will become and the less distracting your presence will be to the employees. You will see, hear, and feel the pulse of the company. Good leaders can foster this communication and still maintain the normal chain of command.

Like all walks of life, some shop personnel will be more conversant, and you will have more in common with some individuals than with others. It is not necessary to converse with everyone equally. You will gravitate to certain individuals, but there will be times where you may want or need to discuss situations with the most reticent of individuals. I have listed some typical ice breakers.

- Are you making any good ones?
- The Steelers took a licking this weekend (or any other team based on clothes or memorabilia in the area).
- How do you like the new machine?
- How old are your kids?
- What's going on with this machine (pointing to the one that is not running)?

This situation is no different than any other conversational situation. The point is that during normal conversation with people throughout the organization you will be exposed to good and bad details that otherwise you would have no idea that they were occurring.

STRATEGY EXECUTION

Business strategy requires that the operations be capable of delivering new or custom products at the cost, quality, date, and specified differentiation and service level. There needs to be an operations strategy that is aligned to the business strategy and aligned with other functions such as marketing. Strategy normally does not fail or underperform simply because the strategic plan is poor or flawed. The overwhelming reason why strategies fail is poor execution.

To achieve success, the business strategy needs to be deciphered and translated to an operations strategy. The future operational requirements need to be measured against the current operational capabilities. The "gap" or "delta" between current capabilities and future requirements needs to be analyzed for volume, variety, quality, lead time, cost, capital equipment, and various direct, indirect, and overhead manpower changes. A plan must be developed to close the gap within the necessary time frame. Depending on the number or size of the gaps, the magnitude of the plan(s) may be small or large.

Bigger organizations may have larger scope strategies yielding more complex plans to close the gaps, but the bigger organizations also possess more resources at their disposal. It may be more difficult for the medium or small organization to implement their smaller scope strategy.

Regardless of the size of the organization or the size of the gaps, it is the operations management team that ultimately must translate the business strategy, create an operational strategy, and execute the plan. I am a big fan of the Ben Franklin quote "vision without implementation is hallucination". Yes, companies need people who can get things done.

OPERATIONS MANAGEMENT CONCLUSION

It should be evident that your operations management team is the heart of the manufacturing company. This group is directly responsible for cost, quality, delivery, strategic execution, culture, continuous improvement, and ultimately customer satisfaction. A healthy heart can overcome many other shortcomings. An unhealthy heart will yield diminished returns, regardless of how much exercise or nutrition the body receives. Without a healthy operations management team your organization will yield diminished returns, regardless of its nutrition via capital investment in equipment, design, R&D, or marketing.

There is no other group in your company that directly touches the range of stakeholders from the supply chain to customers as your operations management team. Core competencies can erode, sometimes quickly, with changes in technology and market conditions, not to mention acquisitions and global competition. What if your core competency was your operation management team? I would suggest that this leads to an agile and learning organization that can transition and adapt to changing technology and changing market conditions. People and a culture that continuously improve themselves as well as improve the processes and equipment can acquire and transfer knowledge and insight (Giesecke and McNeil 1999, 158). This core competency allows the organization to meet new challenges, implement new strategies, acquire new customers, absorb acquisitions, and out-pace the competition.

For these reasons it is essential to employ leadership and an operations management team that is talented, incentivized, and motivated.

13 SUPPLY CHAIN EXECUTION

The market trends we have discussed in this book are pushing companies of all sizes to be less vertically integrated and to rely on a supply network to meet the speed and cost requirements of their customers. The supply chain partners are more numerous and more geographically dispersed. Nearly all medium and large organizations have established sources in low-cost regions as a necessity to be competitively priced for either castings, forgings, or basic machining. All these factors increase the supply chain management challenges precisely when supply chain execution is in greater demand.

Shorter product lifecycles coupled with greater product differentiation do not allow time for organizations to acquire and debug new equipment or processes. Environmental and certification requirements extend the development, installation, and debug period. These factors increase the investment, which further drives the ROI to unacceptable levels. Sourcing the new or risky processes to proven suppliers can be an easy decision. The supplier can spread the expertise and cost of capital and compliance across numerous customers, many of which may be competitors to each other, while providing a service not available through vertical integration.

The Council of Supply Chain Management Professionals (CSCMP) defines supply chain management as:

> Supply chain management encompasses the planning and management of all activities involved in sourcing and procurement, conversion, and all logistics management activities. Importantly, it also includes coordination and collaboration with channel partners, which can be suppliers, intermediaries, third party service providers, and customers. In essence, supply chain management integrates supply and demand management within and across companies.

And as:

> Supply chain management is an integrating function with primary responsibility for linking major business functions and business processes within and across companies into a cohesive and high-performing business model. It includes all of the logistics management activities noted above, as well as manufac-

turing operations, and it drives coordination of processes and activities with and across marketing, sales, product design, finance and information technology.

Within the machining industry, there is a wide variation of complexity and dependency between the various types of machining organizations and their supply chain. The smallest and least complex organizations only need to purchase commoditized raw materials. The larger and more complex organizations require acquisition of assemblies, machined components, special processes, services, castings, and specialty materials. All other organizations are in between these two ends of the spectrum.

In general, larger organizations carry higher overhead and burden, which makes subcontracting more favorable. Smaller organizations carry lower overhead and burden and perform as much labor internally as possible. What all the machining organizations have in common is that the chain is only as strong as its weakest link. This is very apropos to a manufacturing supply chain. In fact, the supply chain is frequently the weakest link for quality and delivery.

Many larger organizations have extensive purchasing and supplier development resources to minimize supply base risk and to also improve the quality, delivery, and future capacity of their vendor base. They may have enough resources to imbed personnel within key suppliers and generally execute a sophisticated supply chain strategy. It is not my intention to provide coaching to these groups of professionals.

The scope of supply chain architecture at large companies is end to end or from the supplier's supplier to the customer's customers (Cohen and Roussel 2013, 61).

My focus for machining operational excellence is more narrow. I want to discuss the external supplier relationships needed by machining organizations to provide raw materials, components, machining services, special processes, and critical MRO. Managed properly, your supplier network can be a competitive weapon and will allow you to compete in market segments that otherwise would be out of reach due to investment and learning.

While increasing reliance on your supply base can increase your competitiveness, there are hazards. Less thoughtful leaders can fall victim to "thinning the core" by creating an over-reliance on the supply chain (Cohen and Roussel 2013, 110). Kodak seems to be a prime examples, as there was a focus to protect the market share and profit of the film business, but not the same competitiveness with the camera itself. Consequently, there was not the breadth of talent and resources in the organization to develop and build a comparable product as the Japanese competition during the shift to digital photography.

Strategically, leadership should determine the product or process differentiators along with any proprietary technologies and not subcontract these items. Too much outsourcing can weaken the organization and create a cycle of dependence that reduces the need for internal skill, talent, processes and technical resources. The result is a greater reliance on the supply chain and a continuous cycle, as shown in Figure 13.1.

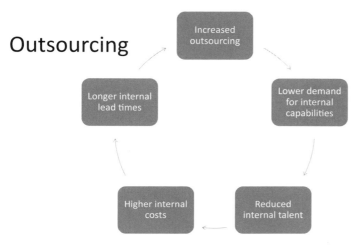

Figure 13.1 Cycle of subcontracting dependency

There are hidden costs to outsourcing that are frequently ignored, leading to poor make/buy decisions. Freight cost and double handling of components can erase savings. A good example is related to Company L that we discussed in the Scheduling and Capacity Management chapter. Machined components were outsourced to India and China. Quality and delivery issues necessitated air freight, which alone increased freight cost by 450K during the first year. Each component needed to be heavily oiled and individually wrapped for the shipment, which was scheduled by ship but was frequently diverted to air. The assembly cell was not set up with an efficient cleaning method to remove the oils and did not include the space for removal of considerable packaging material. Preparing the overseas components for assembly could be up to 40% of the total assembly time. This was not part of the savings calculation. Eventually, the components were sent to a third party for unpacking, cleaning, and repackaging in smaller quantities suitable for the assembly cell. Mechanical properties of the material as tested overseas did not match the material properties as tested in the U.S.

In aggregate, these unexpected but not surprising complications eliminated any cost savings projected by Company L. Quality issues, late deliveries and expediting efforts were extensive. The freight and additional handling costs were invisible, as they did not affect margin but were buried in operating expenses. Similarly, travel costs for engineering, purchasing, and quality did not show in the margin or in the project cost savings. The indirect time from these sources were lost opportunities to improve the business elsewhere.

Machining organizations need to consider their primary machine tool distributor(s) as part of their supply chain. It is not uncommon for the manufacturer to be out of business, the particular mechanical or electronic component required for a repair to be unavailable, or to experience long delays for spare parts on newer equipment. It is prudent to utilize your leverage on orders for new machine tools to guarantee local and domestic stock of spare parts and service. And it is prudent to consider longevity and stability of the manufacturer of the machine tool as well as the stability and supplier of the manufacturer of the machine control.

BEING A BETTER CUSTOMER

I can recommend one guaranteed method to improve the performance of a supplier: look into the mirror! The first step to having better suppliers is to be a better customer. When I sit down with the leadership and operations folks from a supplier, it never fails to highlight opportunities which we can work together to not only improve the supplier performance but also resolve root causes and reduce stress. Better understanding the supplier's business structure, customer base, operation flow, and how my business fits their model begins a foundation for improvement. Understanding what percentage we are of the suppliers' total business or profit is crucial. It sets the stage between begging, asking, and telling. It helps to determine whether you need them more than they need you—which is always important information in a business relationship. It is not uncommon to discover that as a customer we have been a little demanding and non-communicative. The daily grind of business sometimes becomes too impersonal. Emails and phone calls will request extreme expedites, constant demands for updates, and overall price and quality pressure. Many of these situations require a value stream analysis to determine why the flow of product has become so difficult and actually create a level-one root cause corrective action.

It helps to explain to the supplier their importance to your business and the importance of the supplier to your customer. We all lose sight of our impact to the end customer, and the further down the supply chain the greater the disconnect. Meeting and thanking some of the supplier's people pays dividends not only in the short term but also in the long term. Overall, providing better schedules, lead times, and communication can solve many issues. You may still need some rush orders, but not all orders should be a rush.

Many of the keys to internal success are also the drivers for external success within the supply chain. Design for Manufacturability and standardization benefit everyone. When the design engineer reuses existing components, success will be higher. Why alter a casting pattern when your machining operations can implement the new design through machining? This will eliminate cost, lead time, and considerable risk from the supply chain. If manufacturing engineering standardizes on material sizes, you can purchase and stock much less variety of material sizes. This allows your raw material vendors to order in higher quantity, stock at their location, and deliver as needed. Because your variety is reduced, inventory can actually be lower, and inventory turns may be higher for everyone. Custom packaging or returnable dunnage prevents handling damage to, at, and from your special processors. All of these initiatives are examples of being a better customer.

SUPPLY CHAIN CONCLUSION

Leadership determination of which products to outsource and which products not to outsource, accompanied by which processes you will do and which processes you will not do, is the first step to aligning your supply chain strategy with the business strategy. You should have basic supply chain metrics to not only understand whether you are meeting your objectives, but which suppliers are trending in the right direction and which may be in jeopardy of a severe quality or delivery setback. Many of the highly expensive auto recalls are supply chain-related. Suppliers occasionally exit the market with little or no warning. This may be due to financial failings or shifting to a less regulated or less demanding industry. I have experienced a critical aerospace casting company exit the aerospace business for the higher-volume and less regulated truck/auto market. This created extreme supply chain disruption. I have also seen an aerospace machining supplier close the business with no warning. This created havoc for a very large customer, which in reality spread the havoc to the rest of the customer's machining supply chain. Similarly, a friction paper supplier that was destroyed by fire was the sole source for several transmission components. Some of these predicaments are more unexpected than others, but it is mandatory that your organization maintain personal contact with suppliers and manage the right metrics to avoid the cataclysmic scenarios.

If you are required to subcontract unique processes such as specialty heat treats, coatings or testing (x-ray, ultrasonic, mag particle, etc.), you may find that delivery or quality of critical items were executed superbly internally but negatively affected by your supply base. Determining the best partners and maintaining good relations through communication is an integral part of your success. Overall, two machining companies of equal internal performance will be separated in total performance by their supply chain. Continuous improvement yields the same ample benefits in your external supply chain as in your internal operations.

LAYER 4
RESULTS

14 LOW SCRAP/REWORK

The presence of scrap or rework or the absence of scrap or rework is the "result" of the effectiveness of your quality system, people, equipment, CNC program validation, and all the other building blocks structurally supporting scrap/rework on the Machining Pyramid. Every dollar spent on scrap and rework is subtracted from the profit as demonstrated through the hierarchical placement on the Machining Pyramid. Similarly, scrap and rework require equipment and personnel to replace the damaged product, which will result in lower on-time delivery, expediting cost in the form of overtime, shipping, and outside process premiums, and schedule disruptions to accommodate the repair and replacement. In turn, each of these will result in immediate reduction in profit or a long-term reduction in profit. Scrap and rework normally generate late deliveries by the time the repair or replace activity can be completed. The repair and replace activity itself will consume capacity needed to ship other orders on time. Essentially, each scrap or rework event can result in two late deliveries. Worse, repair and replace can create a spiral of late orders and dissatisfied customers leading to less orders and lower revenue.

Common sense dictates that scrap and rework will wreck customer satisfaction, delivery, and profits. It is safe to state that the scrap/rework elimination is a driver for success. Further, when you discover scrap or rework, an analysis should be made to find out whether any of the non-conforming product have been sent to the customer. The more scrap and rework, the more statistically probable that some has not been contained. The historic knowledge that inspection identifies approximately 85% of errors is still reality. It is inevitable that when processes generate non-conforming material, some will eventually leak through and be found at the customer. It is evident by millions of automotive recalls each year that even the best Japanese manufacturers are victimized by latent quality defects. Machining components possess more geometrical features than other types of manufactured items (stampings, molds, weldments). Dies and molds will produce tens of thousands or millions of parts before replacement. There is no wear and very few process inputs to assembly. Every machining operation contains a few to dozens of cutting tools that are immediately wearing out and compromising the ability to meet print specifications. Some cutting tools remove material from both sides of a part which will double the effects of the worn tool. Components cut in a 4th axis will also double the error. Since a machined component will have dozens and perhaps hundreds of dimensions, the requirement to control all inputs is so severe that there will be a certain amount of scrap and rework.

Like other key indicators you need to establish metrics to determine trends and to provide the ability to drill down. Common scrap/rework metrics include total dollar value, scrap and rework as a percent of revenue, and scrap and rework as a percent of value added. Value added is revenue minus materials, subcontracting, and purchased components. Value-added metrics enable comparisons between facilities with a different level of purchased content and comparisons of the same facilities that have variations from month to month with a magnitude increase in purchase content, subcontracting, and/or product mix. In other words, percent of scrap/rework to revenue is meaningless, if one quarter of a company has 3M in purchase content and the next quarter the same company has 6M in purchase content.

COST OF QUALITY

The other common metrics are the total cost of quality and the total cost of quality as a percent of value add. Cost of quality is the sum of warranty, scrap, rework, prevention, and appraisal. Some organizations also include training.

While it is never desirable for cost of quality to increase, there are some mitigating factors. The first is the introduction of new products. There will invariably be some learning curve and debug required. If the amount of new product flowing through the company is a higher percentage or of a higher complexity, there may be a corresponding movement upward in the cost of quality metrics. It is not practical to expect new processes to be as robust and flawless as those that have been fine-tuned over months and years. Clearly, quality needs to be built into the new products.

The second mitigating factor is when growth and/or attrition creates a need for a new shift or a statistically significant number of new employees. Despite the best efforts at hiring and training, it is inevitable that new employees will be a part of the interaction leading to increased scrap and rework. Finally, new processes also require debug, training, and a learning curve. Changes to products, people, and processes may generate expected non-conformance and therefore should pose little or no risk as an escape. The result for some period of time may still be a higher cost of quality.

What is a reasonable cost of quality as a percent of value added for a machining facility? What is a reasonable amount of scrap/rework per month? I can answer this in several ways. First, your goal should not be to be "reasonable" but to achieve operational excellence for your industry and segment. Second, it is much better that you have a high cost of quality driven by prevention and appraisal rather than from scrap, rework, and warranty. Since quality is a driver for productivity, a high cost of quality due to prevention and appraisal may be recouped with higher margins. In other words, cost of quality has two components: the cost of good quality such as prevention and appraisal and the cost of bad quality such as scrap/rework/warranty. I have seen metrics that include appraisal as a cost of poor quality, but with machining appraisal is a necessity to control the inputs which yield conforming products. Therefore, I firmly include appraisal within the cost of good quality.

Overall, there are three components that drive your cost-of-quality level:

1. Industry: Medical and aerospace will have a higher baseline than tool and die or a commercial commodity. Prevention will be high due to a more complex quality system and more frequent audits. Appraisal will be higher due to ISO and customer requirements.

2. Industry segment & business model: Each industry contains segments that vary in complexity, variety, and lifecycle. Aerospace interior aluminum hardware is not the same as aerospace Inconel engine components. Likewise, an orthopedic contract manufacturer may focus on high-volume cervical plates, while another orthopedic contract manufacturer may focus on R&D, low-volume and patient-specific cervical plates and spinal products. These differentiating niches within the segment should create a different cost of quality.

3. Execution: Some organizations have implemented and practice the Machining Pyramid fundamentals and will incur minimal costs of poor quality for their industry, segment, and niche. Some organizations are missing or are inefficient at entire Machining Pyramid building blocks, and this will be displayed in the "results" sections of the Machining Pyramid, i.e., Scrap/Rework, and will lead to a much higher cost of quality.

The majority of machining organizations will be in a range of 7–13 % for the cost of quality as a percentage of value added. If your COQ is higher, you may have ample opportunity for improvement!

In general, robust processes with a higher Sigma level require less appraisal and generate less cost of bad quality (scrap, rework, warranty). This assumes that your business model provides repeat work with some volume. Appraisal cost can be reduced by any business model through increased inspection automation (spindle probe, CMM, in-process gaging) or moving appraisal internal to cycle time.

How important is cost of quality? Two identical contract manufacturers with an annual revenue of 25M can have a 3 % difference in cost of quality—because one practices the Machining Pyramid and one does not; this equates to an additional 750K in profit. This is the short-term direct effect on profit. I have already discussed the long-term negative indirect effect due to increased quality escapes, lower on-time deliveries, and lower or contracted business growth.

Scrap and rework are indicators of other organizational and managerial problems. There is always a root cause. If you are not getting to the root cause, you are not asking "why" enough times. This chapter is not going to tell you how to remedy your scrap and rework problems. If you do not know, start reading again in chapter one!

15 ON-TIME DELIVERY

Most people do not realize that customers have customers. Rarely is a machined product sold directly to the end user. Nothing will affect a business relationship more than a significant late delivery to your customer. Why? Only bad events occur when your customer does not receive the product on time. First, you may be idling your customer's workforce, which will also prevent your customer from shipping to their customer. If their work force is not idle, they may be forced to work on secondary priorities. Second, as I have been told several times, "your $500 part is preventing me from shipping a $40,000 assembly which is preventing my customer from shipping a $40,000,000 aircraft". Instead of having the attention of just purchasing and supply chain management you now have the attention of the entire organization of your customer. Third, most large organizations award new business and eliminate suppliers based on a scorecard that incorporates a combined delivery and quality rating. A poor delivery score for an extended period will either get you phased out slow if you are an important supplier, or booted out quickly if you are not a strategic supplier.

How do you measure on-time delivery? How do your customers measure on-time delivery? Chances are that you have customers with different measurement criteria, and these may also be different from your measurement criteria. It is very important to establish a consistent internal metric to not only measure trends but also measure effectiveness of people, projects, and preventive actions intended to reduce lead times, reduce bottlenecks, improve scheduling, and improve macro- and micro-capacity.

The various measurement techniques for on-time delivery are listed below:

1. The percent of pieces delivered on time. This is a simple and accurate method, but if your product mix includes some low-dollar parts of higher volume and some high-dollar parts of lower volume, you may have a skewed metric. One order of 500 pieces would equal 500 orders of one piece using this metric. It is common that 500 pieces might equal only one low-volume piece in dollars.

2. The percent of line items from the customer's PO that are delivered on time. This sets each line item as equal regardless of volume or dollar value per line

item. In this method, partial deliveries are considered late. Many customers request multiple deliveries of the same part number within a short time period to reduce their inventory. The trap with this scenario is that you may incur a single root cause for a late completion of your production order that generates multiple late line-item deliveries.

3. The percentage of dollars delivered on time. This metric provides credit for partial shipments, and it provides balance between the high-volume parts of a low price and the low-volume parts of a high price.

It is wise to measure your "on time" identically to your major customer(s). If you have multiple major customers, choose one or two methods for your internal metric, be consistent, and graph the trends.

There may be a gap between your internal metric for on-time deliveries and your customers' measurement of your on-time deliveries. One reason is the definition of "on time". As a supplier we record the line item as shipped when the shipping department performs the shipping transaction. Your customer is recording the measurement when his receiving department performs the receiving transaction. There can obviously be a gap which will alter the on-time metric between the two organizations. A customer's PO will state the delivery date. The easiest method is to enter the delivery date into your ERP system one or two days prior to your customers delivery date based on known shipping time.

Packages may not actually leave your building the same day as the shipping transaction, and there also may be a gap of one day or more between the time your customer receives your package and the time the customer actually performs the receiving transaction.

If your customer places considerable weight on their on-time metric you may want to obtain a monthly list of all transactions to insure the metric is accurate. There are many ways that your customer can negatively affect your ability to ship on time. Spec changes, ECNs, PO changes (quantity, date), print errors, approvals, or failure to timely supply agreed-upon items such as raw material, gages, models, etc. In these circumstances, you should ask for a PO change with a new date or for the particular line item to be moved from the late category to the on-time category and for the metric to be recalculated. You may find items that simply were not received until well after they arrived, as stated above. Some customers require reason codes why orders shipped late to better understand your trends and to provide reason codes that designate the customer as the root cause.

The final twist is the customer who counts orders late before you even ship! At the end of the month any order not yet arrived is counted as late for the monthly scorecard. If the order has not yet arrived at the end of the following month, it will lower your scorecard yet again.

ON-TIME DRIVERS

We discussed in the Machining Pyramid chapter that on-time delivery is a percentage with a numerator and denominator. There is no single action or small group of actions that can improve on-time delivery. There certainly is no magic dust to improve the on-time delivery numbers. There needs to be a lot of smart, dedicated people grinding it out each day. To improve this metric you need to improve the entire Machining Pyramid. What are some of the drivers for on-time delivery?

1. Who in your organization makes the commitment for the delivery date? Is it the individual salesman, plant manager, team, or other? Customers frequently pressure suppliers for a delivery schedule that is not realistic. The sales person is the most likely to give in to the pressure. The sales person should be selling quality and reliability and not committing to unrealistic dates. Aggressive delivery schedules should be a team or operations decision. Generally, it is only the operations management team that comprehends the total scope of the project. The total scope includes fixture design/manufacture, CNC programming, metrology, scheduling and capacity. Additionally, what are the risks and what other projects will be competing for scarce resources? For these reasons key delivery dates need to be left to operations. A realistic schedule is the first step to controlling your on-time delivery.

2. Scrap/Rework: We have obviously talked a lot about all the variables, people, and culture that need to be optimized to limit the amount of scrap and rework. Remember, the easy work is in the low-cost countries. Suffice it to say that scrap and rework are a driver for on-time delivery. Whenever scrap and rework occur it is probable that the order will ship late and the snowball of jamming your capacity for replacement parts will generate more late orders.

3. The least expensive and most flexible type of inventory is raw material. Raw material availability will improve your on-time metric for several reasons. In some instances material availability will secure the order with a premium, as the customer wants to move fast and understands you will probably be the quickest to deliver because of your head start with the raw material. Choosing the type and size of material to stock needs to be an operational strategy. Not only will this eliminate the waiting period for new material to arrive, but this can also be a buffer when the unfortunate scrap occurs. This can create a large savings, if it prevents equipment setups or outside processes from occurring twice, because the replacement material is readily available. Even the best organizations have some degree of scrap. Material can be available while still maintaining acceptable inventory levels. There are means to utilize common sizes of materials to reduce the variety that needs to be purchased and inventoried. There will also be expected price breaks for ordering larger quantities. For smaller parts, not cutting full bars or splitting sheet stock at the distributor will generate a savings and provide buffer material. Again, raw material inventory

should be strategic, based on history, forecast, cost, scrap risk, setup times, and lead time requirements.

4. Is your supply chain predictable? The vast majority of machining organizations possess minimal leverage with suppliers. Leverage is the friend of the big organizations with higher volume (see Figure 15.1). Everyone else learns to beg and plead, and particularly during times of a strong economy. With the strength in aerospace during the last five years I have witnessed even small coaters and casting houses occasionally defy the leverage of large companies and dictate both price and deliveries. This is possible when the purchased item or service is lower-volume and complex, or when it is lower-volume and the number of approved suppliers is limited. The situation for midsize to smaller machining companies that require certified materials and components with limited sources is particularly challenging. Even worse, large customers dictate to their supply base where to obtain specialty processes for heat treat, coatings, and testing. This insures there will be little or no leverage, as the processors comprehend that they have little or no competition. It is not uncommon for some production orders to spend more time outside of the facility than inside of the facility during the total manufacturing cycle. It is a hard pill to swallow when you get verbally whacked and your scorecard penalized by your customer, only because the supplier they dictated for you to use for a specialty process was two weeks late. Likewise, your customer may dictate where to buy castings or forgings, which will also be problematic. Needless to say, much effort, planning, and praying needs to be involved, if your supply chain is going to be a positive factor rather than a negative factor to your delivery metric.

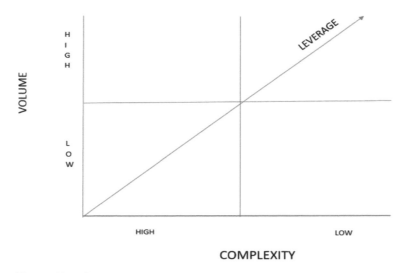

Figure 15.1 Supply chain leverage matrix

5. Prior to committing to delivery dates can you look into the future? I do not mean a crystal ball or a Ouija board but a capacity management tool that displays the future demand at future dates on the same workcenters which are required for the delivery dates you are contemplating. If you have created such a tool, you will find that sales may resolve the issue on their own without pressing for a team or plant manager decision. It is common for your capacity management to demonstrate clear conflicts between existing orders and proposed orders. With firm knowledge and data, the sales person will work with the customer to obtain a realistic date or together with the customer reschedule existing orders to a later date to create capacity for the proposed order. There is a time lapse from the initiation of the quoting process to the point in time sales requests a firm delivery date from operations. During this time lapse the customer is probably creating a project schedule and providing commitments to their customer. Sales should provide the customer realistic delivery guidance early in the process versus tacit approval through silence or ambiguity. The further along the customer's customer is in the project planning process, the more difficult it becomes to accept a delay in the schedule. Knowledge is power, and in this case, knowledge of future capacity by workcenter is the power to improve your on-time delivery.

6. Despite good planning and best efforts, there will be times that you have more work to be performed in a given time period than your capacity allows. This may be at the macro level or at the micro level in one or more workcenters. The reasons may be as simple as an underestimation of hours on a large project, downtime on key equipment, or you quoted a large number or projects expecting a 60% hit rate and instead were rewarded (or penalized) with a 90% hit rate. What now? The answer is what I refer to as "surge capacity". This would be Lindberg's reserve gas tank on a very cold night over the Atlantic. Every pilot (plant manager) or navigator (production control) appreciates some margin of safety other than a parachute. Surge capacity can be generated in several manners, but the roots lie in the Machining Pyramid. In other words, surge capacity needs to be part of your daily operation, culture, and planning. If you wait to try to create surge capacity until you actually need surge capacity, you should just reach for the parachute. In fact, if you have not created surge capacity, then, like the pilot, you do not even have a parachute in your plane. Below are surge capacity fundamentals:

a) Avoid single-sourcing purchased items. If it is not practical to dual-source, then rotating the sourcing is the next best alternative.

b) Avoid single-sourcing internally manufactured items. Stated in different terms, do not have just one piece of equipment that can perform a type of machining or process. If this is your current situation, find an outside source and get them qualified. At a minimum, you should insure that your single piece of equipment does not contain obsolete components, that the PM (preventive and hopefully predictive) is exceptional, and that you have spare parts available in-house or in stock at your local distributor.

c) Outside sources as overflow for all internal non-proprietary processes. Re-gardless of the root cause, once you are behind at a workcenter, it can be a difficult journey to get caught up and back to completing orders on schedule. In the meantime, the late deliveries can be significant. Worse, this can be a contagion, as all downstream processes will need overtime and expediting, since orders are now past due.

d) Cross-train people to move to the workcenters in most need of surge capacity. Not only do the people need to be trained but they must periodically be al-lowed to operate the equipment to retain their skill and knowledge. This is more easily said than accomplished. Most machine tools are complex, are equipped with a variety of CNC controls, and the processes may be very dif-ferent. This cross-training and skill retention needs to be part of your culture and rewarded with praise and compensation for both the machinist and his manager.

e) When you cannot move the people to the work, you can move the work to the people. When there are no extra machine tools or shifts available to move people to the work, then it may be possible to re-process the work onto a type of machine that is available. Again, your operations management team should have a surge capacity plan for all workcenters, and this re-processing should be tested and qualified in advance.

f) The last and perhaps most important surge-capacity fundamental is staffing levels and use of overtime. If your normal mode of operations is maximum overtime each week, then you do not have surge capacity for high workloads nor for people and equipment losses. Conversely, if your average hours per week range from 45 to 55, then you have some degree of macro surge capac-ity to increase hours until you can implement one or more of the plans above, or until equipment and staffing is permanently augmented. Again, this approach needs to be part of your culture and DNA. When the economy is fa-vorable and business levels are acceptable, there is no reason to not be ahead of the curve with the staffing level and training.

7. New orders need a formal or informal review to determine, if they need to be assigned some type of project management. In general, there are two classifica-tions: repeat orders and new orders. New orders that are very similar or in the same part families as existing business do not receive project management. New orders that are not similar to existing families need to be reviewed for new pro-cesses, new specifications, new part features, tolerances, materials, etc. If risks are identified, an order review team needs to establish a plan to test and elimi-nate the risk. AS 9100 requires a risk analysis on new orders for this reason.

Overall, identifying orders that contain either a quality or delivery risk and assigning the appropriate level of project oversight or management will increase your scorecard. Developing a culture of focusing as much attention on orders at the beginning and middle of their lifecycle as at the end is a pre-requisite for taking control of your deliveries. I cannot stress this last point enough.

ON-TIME STRESS

The effort to ship products on time to customers places more stress on individuals and groups of employees than any other business activity. You will most probably not have had a customer threaten to fire you because your margins on his products were 5% too low, but I will bet that you have had threats over delivery. Profit is required, but a few less points of margin are not noticed until the end of the month or quarter. Profit is a function of many variables such as materials, overhead, and design. Profit can be gradually improved, providing that you do not get fired by the customer first due to late delivery. It is common that poor execution of scheduling, capacity, and project management leads to sacrificing margin at the end of a project. Profit can be increased via new processes, tooling, or equipment. There is no definitive answer to how much profit should be obtained for a project. Conversely, on-time delivery is black and white. On-time delivery is win or lose. With on-time delivery there is a line in the sand. On-time delivery frequently depends on the last few manufacturing operations working long hours at a hectic pace. This scenario can be repeated daily. Your sales executive or sales account manager will pressure the entire operations group to keep the project moving. The sales department may have internal conflicts with each other over which projects are given priority in engineering, programming, manufacturing, and inspection. Customers can and do ask for detailed progress reports, milestones, and regular updates. When the customer does not receive satisfactory commitments from sales, they can and do contact plant management or higher levels in the organization. The stress flows through all levels of the company!

Leadership needs to create systems and an organizational structure that permit the exceptional effort required to ship on time to become more routine. Addition of a second shift shipping person to receive late deliveries and take last-minute deliveries to the airport or to FedEx is simple but effective. Management must eliminate the fire drills and throwing overtime at projects to push them over the finish line. Our people are not hamsters in the wheel to always run faster—that is not management and it is certainly not leadership. Use the Machining Pyramid tools and reduce the delivery stress through planning and execution.

LAYER 5
ULTIMATE RESULTS

16 PROFIT

Some may argue but I would submit that profit is the second most important result of the organization. If you lose your customers, there will be no chance of a profit, hence maintaining your primary customers must have precedence over profit for any individual decision or short period of time. A low profit, or even a loss for a short period, may be reversed, if you maintain your customer base. So why is profit sitting atop our pyramid and not customer satisfaction? The answer is that many times there are companies with satisfied customers, but no profit or not enough profit to maintain the organization during the course of normal economic cycles. There are many factors that contribute to this predicament, but the most common are that the organization does not understand its own cost structure, a lack of efficiency compared to competitors, and the inability to say "no" to a customer or to say "yes, and the new price is …". The company management's primary task is to make money and keep it, whether for dividends, equipment, facilities, or R&D.

Profit is at the top, because this book is for the machining organizations and not the customers. Profit is at the top of the Machining Pyramid, because it is the ultimate result, the ultimate indicator, and the ultimate metric of how well machining companies are managing their business. Profit is certainly a byproduct of the Machining Pyramid. It is a lagging indicator that can only be improved from the bottom. There may be occasions that you have a dominant market position due to a proprietary design or process, which will create strong profits even though inefficient processes or management are leaving money on the table. And there may be circumstances of a strong economy, or strong activity in a certain industrial sector, which creates a capacity crunch enabling you to increase prices. Primarily, long-term trends in your profit margin are indicative of your actual performance.

There are several accounting variations on profit, but I want to keep it simple and measure what is normally termed gross profit and gross profit margin. Gross profit is shipments minus cost of goods sold. Since the cost of goods sold is the direct labor, direct materials, and overhead to produce the product, the formulas are:

$$GROSS\ PROFIT\ =\ SHIPMENTS\ -\ (DIR.LABOR\ +\ DIR.MATERIALS\ +\ OVERHEAD)$$

GROSS PROFIT MARGIN =

$$\frac{SHIPMENTS - (DIRECT\ LABOR + DIRECT\ MATERIALS + OVERHEAD)}{SHIPMENTS}$$

EXAMPLE:	SHIPMENTS	=	$1,000,000
	DIRECT LABOR	=	250,000
	DIRECT MATERIALS	=	150,000
	OVERHEAD	=	300,000

$$GROSS\ PROFIT\ MARGIN = \frac{1,000,000 - (250,000 + 150,000 + 300,000)}{1,000,000} = .3\ OR\ 30\%$$

A decrease in your gross profit margins over a period of time may be driven by several factors or a combination of factors.

- A lower sales price for your product – this can have several root causes ranging from poor market conditions over increased competition to higher volume. To some extent, this should come as no surprise. The customer is always attempting to purchase at a lower price. Whether you have the leverage to maintain your pricing may be a marketing and engineering function. Regardless, operations should be driving out cost faster than the customer can drive down the price. This is manufacturing, and this should be accomplished. Yes, margins can increase in times of price declines.

- Higher production costs – this includes more hours required to process the order, higher direct labor rates, increased overhead, and/or higher material costs. Organizations with higher overhead ratios are particularly subject to higher burden rates due to the change in the direct/indirect ratio.

- Cost of poor quality through high scrap, rework, and warranty.

- A change in product mix – this means you are selling a higher percentage of lower margin products and a lower percentage of higher-margin products. This could indicate that you are being outfoxed by a competitor. It can be caused by a lack of understanding of the cost associated with an order at the time of quoting, or poor data regarding performance on prior orders for the same part number. There is a saying that there are two types of companies that go out of business: The first is a company that does not know what their costs are, and the second is a company that knows what their costs are and does not do anything about them.

A healthy gross profit margin is necessary for a company to invest, grow, and save for poor economic cycles. There are many machining companies that can survive during strong economic times, but are not financially prepared when the economic cycle turns negative.

What is an expected gross profit for a machining organization? Since the gross profit margin incurs several more deductions on the income statement before net income is derived, it needs to be north of 20%. In general, 20–35% is the range of most machining companies, particularly contract machining companies. Organizations producing commodity type items with no control over their design or distribution channel will be closer to the 20%. Companies with design control over products or processes that cannot be replicated without significant investment can be over 35%. In general, the lower-margin companies have lower SG&A, and hence more dollars flow to the bottom line.

It is best to review your gross profit for each shipment and in the finest resolution possible. If you ship an assembly you should know the expected cost of the assembly (standard cost or quoted cost), the actual cost, and the sales price. You should also be able to drill down into the assembly to review the same data for each individual part. Likewise, if you are shipping individual parts, the gross profit report should show this data for each line item shipment. A weekly report that displays each shipment during the previous week, all the data, and the corresponding percentages is a required tool for driving profit (see Figure 16.1). The responsible parties in the organization need the visibility to which orders were losers or missed the intended margins by a meaningful percentage. Obtaining the data weekly permits analysis before the data becomes stale. When a loser is identified, an analysis needs to be made to determine the root cause. If the root cause was corrected, or will be corrected, then you can accept future orders at the same price. If there was no reason for a negative gross profit other than a low price, you should raise the price before accepting a repeat order. In the end, a machining organization needs continuous improvement and innovation to earn an acceptable margin. If you do not get rid of the losers, they will multiply over the years, as wise competitors will not match your losing price. The process of analyzing and improving the margins fosters discussion and continuous improvement. This activity needs to be performed part by part and should become ingrained in the culture of the company (see Figure 16.2). This graphic is the detailed continuous improvement cycle created by the Machining Pyramid, with the gross profit analysis highlighted. The pursuit to purge losers establishes, with all employees, that their performance and accurate labor-data reporting does affect gross profit and does directly affect decisions of the organization whether to accept additional orders.

A word of caution: I have seen repeat orders rejected, because the average gross profit was minus 5%. In reality, the first time the part number was machined the gross profit was minus 35%, and the second time the gross profit was positive 30%. The average is deceiving. Individuals need to take the time to drill deeper and understand root causes and trends. It helps to have cost data that is accurate, accessible, summarized, and yet formatted for deeper analysis. This is part of converting the black hole into a wormhole.

Gross Profit Report

Week	2/5/20xx

Product Group	Part #	Qty	Revenue	Standard Cost	Standard Cost Margin $	Standard Cost Margin %	Actual Cost	Actual Profit $	Actual Margin %
Aero Cell 1									
	123456	10	$ 1,000	$ 750	$ 250	25%	$ 680	$ 320	32%
	123456-YTD	50	$ 5,000	$ 4,000	$ 1,000	20%	$ 3,900	$ 1,100	22%
	987654	15	$ 1,400	$ 1,200	$ 200	14%	$ 1,500	$ (100)	-7%
	987654-YTD	50	$ 6,000	$ 5,500	$ 500	8%	$ 7,500	$ (1,500)	-25%
	789878	12	$ 1,000	$ 780	$ 220	22%	$ 725	$ 275	28%
	789878-YTD	50	$ 5,200	$ 4,500	$ 700	13%	$ 3,900	$ 1,300	25%
Weekly Totals			$ 786,147	$ 633,124	$ 153,023	19%	$ 621,479	$ 164,668	21%

Figure 16.1

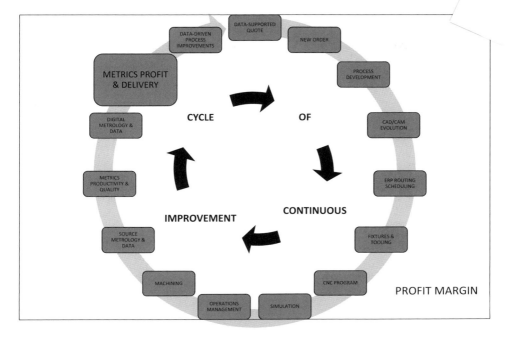

Figure 16.2 Part number specific – data driven – continuous improvement

PROFIT CONCLUSION

The CFO will provide a number of financial metrics in addition to gross margin that will display the direction of the organization. These will include Return on Equity (ROE), sales per employee, Operating Profit Margin (EBIT), Return on Assets (ROA), liquidity ratios, Net Profit Margin and probably others.

I believe that the gross margin is the best measurement that displays management's ability to squeeze maximum profit out of every dollar of sales. Without a strong gross margin, the other profit measurements will not be favorable, and the organization will not have to decide between capital investment or dividends. It is exactly for this reason that the Machining Pyramid was designed to formulate robust and innovative processes directed by world class operations management that generates low cost of quality and high productivity, all integrated with rich metrics and data that can be mined for costing and pricing that will yield strong gross profits. This formula is repeatable and scalable within one facility or across multiple facilities.

17 FOR MACHINISTS ONLY

I want to address three separate groups of machinists: those that have not yet entered the profession but are either currently considering or are already enrolled in an educational training program, those that are in the early stages of the machining profession, and the veteran machinists.

CONSIDERING MACHINING

Why should you become a CNC machinist? CNC is an acronym for computer numerical controlled, and in reality you are close to flying a jet—processing speeds on steroids, 2000 inches per minute rapid, lightening acceleration, unmatched precision, top secret algorithms, sensors, probes, lasers, and more. Everything but the parachute. You can earn great pay and get paid to learn. A machinist learns much, much, more than just machining. You will learn metrology, quality, engineering fundamentals, problem solving, metallurgy, communication, geometry, and math. You will also learn responsibility, self-confidence, and maturity—quickly. You are in control of someone else's jet, your team (fellow employees) is counting on you, and a large customer needs the components that you are machining. Yes, the components you are making will be in a plane, car, tank, ship, or machine tool in a few days or a few months. They may be headed to the military to replace worn or damaged components in a combat zone. Yes, desert warfare leads to rapid degradation of all military hardware. I have had Air Force senior officers tell me that they have aircraft grounded due to MICAP (Mission Impaired Capability Awaiting Parts), and the enemy knows when there are less planes and helicopters in the air. The components need to be correct. But do not be intimidated. Becoming a machinist is a gradual process. Your employer will provide training at a pace you can absorb. If your parents want you to be mature, this is the fastest path.

If you become a machinist and decide later to change your career choice, you will have a proven skill set that is marketable to either your current or future organization. If you combine your machining skill set with future part-time or fulltime education, you will be heads above your peers. As a machinist you will have exposure to critical components in the industries already mentioned, or perhaps to medical instruments or medical implants. You will be interpreting blueprints, utilizing precision-measuring instruments, performing math and geometry, and you will be in command of sophisticated equipment. Some apprenticeship programs partner with college classrooms to earn an associate in engineering at no cost (see Grob Industries).

The most rewarding and challenging management or sales positions are those in a technical field requiring a technical background. With a machinist skill set on your resumé you will have achieved the most challenging competency requirements on most job descriptions.

I am sure that you have heard of the "baby boomer generation". According to Forbes, 60 % of American machinists are over 45, and 22 % are over 55. These senior machinists are not only the most highly skilled but the most highly paid. Likewise, machining facility management and support personnel are rapidly retiring at a high rate, as these coveted positions tend to have a higher average age. These positions include manufacturing engineers, CNC programmers, quality engineers, inspectors, maintenance, supervisors, and plant managers. Many of these support positions will be filled by promoting existing machinists and will create even more vacancies for new machinists.

So if you are considering the machining profession, there will not only be openings but there will be openings for advancement. Anecdotally, those of us who work in the machining industry are experiencing more of a vacuum than an "opening". I expect the shortage of skilled machinists to become more severe. The combination of replacing support personnel, coupled with normal attrition due to health reasons of an aging workforce and retirements, will decimate the skilled machinist workforce.

How do you find a good machining organization, obtain employment, and become a successful machinist? First, you have to decide whether you want to begin training for a machine operator or a machinist. A machinist is more difficult and will be more highly paid. If you begin as a machine operator, you can make the transition to machinist sometime in the future, if and when you feel ready. A machine operator performs no setups or only straightforward setups. A machine operator will not debug programs and is generally instructed to seek help, if any problems arise. Conversely, a machinist performs many types of setups, debugs new processes, and solves problems.

Organizations with high-volume products generally employ primarily machine operators. Design and build companies and machine shops generally employ primarily machinists. Your resumé and/or your application should stress your background with computers, software, problem solving, math, and dependability. You should indicate that you are flexible to work all shifts. Doctors, nurses, policemen, etc. all work many different shifts for many years. It is common for organizations to offer shortened workdays, work hours, and higher compensation for nightshift employees. Frankly, with DVRs and weekends off, many people enjoy the nightshift, and some with small children may save substantial daycare expenditures.

During your interview, you need to display the normal good interview etiquettes. The interviewers want to know that you are trainable, reliable, and a team player. As you are asked questions, take the opportunity to stress these facts as part of your answer. Always have a short list of questions ready for the end of the interview or for when you are requested to pose your questions. Below are some generic but reasonable questions:

- What type of machining do you perform?
- What brand of equipment do you utilize?
- Does your company offer tuition reimbursement?
- Can I be trained on more than one type of equipment?
- What can I study or learn at this time to prepare me for when I am hired?

As an interviewer there is nothing worse than asking the candidate what questions he/she may have and receiving the response "none". A close second to no questions is questions only about vacation, time-off, and attendance policy. This raises a red flag. Wait until you have the offer, and if the vacation issue is not clear, ask or negotiate at that time. Remember, you have to catch the fish before you can throw it back. I recommend accepting the offer from the best organization, even if it is less than another company. You should always look at the complete benefit package, including healthcare, retirement, overtime, and career development. It is best to calculate total compensation over a three- to five-year period. Always take advantage of a 401k opportunity, as you will never have a better opportunity for return on your investment when you calculate the tax savings, company matching, and dividend yield.

Once hired, you need to make an extra effort to be respectful and thankful to the machinist assigned to be your trainer and mentor. He will be a well-respected machinist, and many people will ask him about your performance. Do you want him to show you the minimum, or do you want him to show you the advanced methods and the full tricks of the trade? Do not hesitate to say "thanks". Do not hesitate to get him a cup of coffee or an article about his favorite team. You need to develop a strong professional relationship with your mentor. He may have trained several people, and you want his best effort. You should take notes, take home manuals, and ask plenty of questions. Demonstrate to your mentor that you are serious and that his time with you will not be wasted. You need to be a sponge, learn to read the CNC code, learn new measurement instruments, and volunteer for any additional training or cross-training.

When someone asks your mentor "how's the new guy working out?", what response would you prefer: "struggling", "needs to mature", "too early to tell", "going to be good", or "he has the 'right stuff'"? Your mentor will be asked this question many times, and it is you that has the power to shape his answer. Your peers and management will form impressions on your progress and long-term potential based on the feedback from your mentor. You cannot change this fact, but you can change your mentor's response by insuring your effort and progress.

EARLY-STAGE MACHINIST

For the machinists in the early stages of their profession: Take a minute and look around. How many people at your company will be retired in 10–15 years? Which department, product, or technology is growing? Set some goals on what you want to achieve. I will advise you though that wanting is just the first step. Many people will "want" the same raise, want to operate the new machine, or will want the promotion. You need the combination of education, training, or performance to demonstrate that you are capable and qualified for the increased responsibility when the opportunity arises. Finally, it is not enough to be qualified. You need to see that you are the "most" qualified. Even though there will be substantial openings for good positions within the machining industry, there will always be some competition, whether it be internal or external candidates. Do not wait until the job you covet becomes available to start building your resumé. This is the most common barrier to achieving goals, so I will repeat myself: Do not wait until your dream job or the position on the ladder to your dream job becomes available to start acquiring the experience and education required to be the most qualified candidate.

Many talented machinists prefer to use their ability to be assigned the most complex equipment or the most complex components. They become group leaders or trainers. They learn multiple machine tools, shop floor programming, spindle probes, and other advanced technologies. These elite machinists will become the most highly paid and normally will have the opportunity to work overtime as they choose. This is an excellent career path with high compensation, high job security, and high mobility. It is not uncommon for these top machinists to earn a total compensation beyond many salary positions within the company.

Many people within machining or within any profession do not seek added responsibility. They prefer to leave after eight hours to go home to family, friends, hobbies, or a second job. These are all good choices.

If you desire a management or support position, always augment your technical education and technical growth with English and writing-oriented classes. Communication with customers, suppliers, and internal personnel mandates effective verbal and writing skills. Many competent technical folks cannot make the transition to management, engineering, or support positions due to the lack of communication ability, particularly writing. When strangers read your emails they only have your words. If you have butchered the English language, sentence structure, and punctuation, it is a problem. Your words also represent your company.

VETERAN MACHINIST

For the veteran machinists, who have been in the trade for some time: I believe that there are two pertinent questions—what is reasonable for you to expect from the company, and what is reasonable for the company to expect from you?

The most obvious expectation you have is that your company simply stay in business. You do not want to move, you do not want to start over, and you do not want to compete with a flood of machinists all looking for the same jobs at the same time. Fortunately, you and your company possess a symbiotic relationship, and you can help the company be successful. Your next expectation is for good wages and benefits. The obvious question is how can your company afford good wages and benefits, when the customers can purchase from low-cost countries with wages and benefits one tenth of what you are receiving? The answer is that your innovation, quality, lead time, and productivity have to be that much higher. If your organization is profitable and you are not compensated comparable to machinists with similar skills in your region, then some of the fault may be your own. You or some of your peers can vote with your feet and leave. When your company realizes the total cost to replace machinists, they will make adjustments. The total replacement cost includes training, scrap, rework, lost productivity, and any lost opportunities. This is capitalism, and it does work—both ways.

Another expectation is for occasional new machine tools, processes, and updates to your facility. This demonstrates management and ownership commitment to your future and a strategy to improve cost, quality, and/or product offerings. With new equipment and processes come new technology and new opportunities. This insures that your skills are up to date and you will be employable, whether at your existing organization or at another company. Normally, a company that invests in equipment and processes recognizes that it also needs to invest in people.

Many employees will have strong opinions that their company should have utilized the investment dollars to purchase a different machine tool, a different type of process, or refrained from upgrading the facility. You may have a valid argument, but the big picture is that your company is investing in the future, regardless of whether everyone is in agreement with the specifics. The facts are that an investment in the facility, equipment, or processes is also an investment in the people. When the combination of people, processes, equipment, and facility are competitive in your industry, you have an insurance policy. If your organization decides to sell your facility, or needs to sell during tough times, then someone will want to purchase your facility. Being desired and being purchased is a compliment and perhaps the ultimate compliment. It reinforces that you are valuable. New owner-ship can be expected to entail new ideas, new investment, and hence new opportu-nities. Facilities being bought and sold is not new and will become more common. Employees have two choices: be valuable and be acquired, or be uncompetitive and be closed.

Most machinists expect or prefer to work some amount of overtime. This is not unexpected considering that five hours of weekly overtime will increase annual compensation by nearly 20 %. Many machinists migrate from company to company, following the overtime like hunters following the buffalo. Predominantly, when the company is busy, the overtime is a win/win for all parties and—within limits—a reasonable expectation.

All workers, machinists included, desire some degree of what has been termed "quality of work life". The characteristics of quality of work life will vary for differ-ent professions. For machining, I expect it to include organization of the work-place, the environment, technical support, and management professionalism. The majority of machining organizations recognize that these factors are critical to attracting and maintaining top talent and have established work environments su-perior to many professions. Some office buildings lack fresh air and have been referred to as possessing a chemical stew. The sick-building syndrome has become a real concern for many office workers. Healthcare, construction, transportation, and many other industries face challenges. On the other hand, many machine shops are now air-conditioned and have a very positive environment, as this not only attracts top talent but is advantageous for the electronics and thermal stabil-ity for machining and measuring tighter-tolerance components.

The remaining machinist expectation that I will discuss is what I refer to as a rea-sonable amount of stress. There will always be some amount of stress, since you are in control of an expensive piece of equipment, the parts on your machine are valuable, and customers are depending on your success. Are you provided helpful documentation, proper measurement instruments, and answers to technical ques-tions? Everyone will occasionally scrap a part and, unfortunately, everyone will occasionally bump (wreck) a machine, so when it is your turn for the inevitable mistake, is the situation investigated and discussed in a professional context?

EMPLOYER EXPECTATIONS

Have you wondered what your company expects from their machinists? Maybe you have been told, maybe you have not been told, or maybe the message has not been clear or consistent. Let us start with learning new equipment and technology. Pref-erably, some machinists need to embrace and master new equipment and technol-ogy. I have mentioned earlier in this book that companies need to automate, emi-grate, or evaporate. You only want the one that starts with an "A". The CNC machine is automation and many new accessories, software, and technologies are other forms of automation.

Your company expects you to be thorough and conscientious. You should not only find the obvious non-conformance but the obscure and minute non-conformance. Quality escapes are expensive and damaging to long-term customer relationships.

Everyone needs to be innovative to maintain a productivity and quality gap above the domestic and low-cost international competition. You know your job better than anyone. Provide suggestions and find better ways to add quality and reduce time. Always remember that two or three singles score a run just as well as a home run. A series of small improvements is just as valuable as a big improvement. You are in competition with many others that want to take your job. Be competitive, keep thinking, stay with it, and win!

Finally, if you have learned the new equipment, been innovative, and managed your time, you are in position to run two CNC machines. In the aggregate, the company's effort and your effort should be able to make it as easy, or easier, for you to run two machines in a Machining Pyramid facility as to operate one machine in a non-Machining Pyramid facility. The competition is moving in this direction, and if you are not, you are falling behind. Ultimately, the machinist has to make this happen. If not, your wages and benefits will suffer.

I mentioned that you and your company have a symbiotic relationship. All your expectations of the company and all the company's expectations of you are mutually beneficial. What a beautiful relationship!

18 SALES

The best sales person (account manager) or the best sales tool is always named performance. This holds true for OEMs, contract machining companies, or machining job shops. When customers receive projects on time, with great quality, and good service, you will receive additional opportunities. More importantly, the customer is likely to pay a premium on future projects for the security of working with a partner that will eliminate risk. Conversely, if the customer receives poor quality and late deliveries, your sales person should be a supermodel, if you hope to get your foot back into their door, not to mention receive any significant business.

Many organizations have a unique approach to accomplishing the comprehensive tasks to effectively provide what is termed customer care or customer satisfaction. Customer care will include different facets, depending on the industry, product, and company size. Quality, delivery, and pricing are one component of customer care, while others are service, training, spare parts availability, responsiveness, innovation, technology, quoting, and reliability. There are simply many approaches based on product complexity, product mix, rate of new product introductions, geographical dispersion, etc. The internal organizational variations include who performs quoting, presence of inside sales personnel, and customer contacts for delivery, invoicing, engineering, quality, and service. The optimum approach for any organization depends on the overall business model and the skill sets available.

Generally speaking, sales professionals at machining-related companies need to be technically proficient. When the customer perceives the sales person and his organization as a resource, a consultant, and/or someone who can solve problems and add value, the sales professional will be substantially more effective. Social sales people, narrators repeating a programmed pitch, and storytellers do not fare well in machining. It is beneficial for the account manager to be strategically oriented or at minimum able to follow the business's strategic goals. Many account managers will make any sale, regardless of margin or disruption to the business. Some projects will require excessive development from resources that are not currently available or require capacity that interferes with existing projects already committed. Worse, some projects that require development will be isolated in scope and not transferable to existing and future customers. Any margin achieved on these isolated sales will be erased by the opportunity cost or failure to strategically develop new products, new options, or new customers.

A strategic account manager will recognize development opportunities that will lead to significant business. Stretching to meet these challenges is acceptable and beneficial.

GROWTH

Most organizations that want to significantly grow revenue simply want to do more of their current activity or processes. This rarely seems to be successful in terms of generating significant growth. More of the same is a formula for 1% growth. Companies seeking to double from 10M to 20M, from 45M to 90M, or from 120M to 240M require a new business strategy. Expanding existing offerings, acquiring new customers, expanding to new markets, and adding more value to current customers is more realistic. This may or may not necessitate reinventing the company business model. Adding engineering services, assembly, packaging, refurbishment, or other services will provide customers more value-added activities upstream or downstream to current offerings. A strategic acquisition to acquire new customers to cross-sell existing products is a viable alternative or supplement to organic growth.

Structure of the sales channel is another method to achieve growth. Some organizations rely upon distributors or sales representatives. Many organizations employ a mixed approach of direct sales and distributors. History has shown that it is not necessarily the OEM with the best engineering and best product that grows but the most effective sales and marketing execution. Creating a dynamic distribution channel supported by a marketing campaign to develop awareness, interest, and leads is powerful. Regardless of the sales channel, it is fundamental to remember that the end customer or the end user is the generator of the revenue stream!

While the goal may be to double revenue, the corollary goal should be to not double the overhead. Rather, it would behoove the organization to double the value of a sales order from X to 2X. The development costs for engineering, fixturing, tooling, gaging, and programming are normally identical, whether the sales order is X or 2X. This goal requires the account manager and the organization to attack larger projects or larger programs. An example would be to shift from machining an aerospace pump housing for a customer's spare parts program to machining an aerospace pump housing for a new commercial program with a monthly volume cadence of 5X of the current spare program. This does not mean that the spare parts business needs to be forfeited, but rather that the expertise and performance may be leveraged for a much larger program.

MANAGING THE CUSTOMER

How does an account manager achieve customer satisfaction or customer loyalty while fulfilling his fiduciary responsibility to his employer? Are there situations that require the account manager to say "no" to the customer? Some organizations and some account managers never, or rarely, say "no" to the customer. The bigger question is whether the account manager is successful at managing his customers. The account manager does not need to say "no" to be effective. He may say "yes, we can do that—here is the new date", or "yes, here is the expedite charge".

The customer, in essence, is a person or a group of people (and occasionally the customer is the proverbial 800-pound gorilla) who have human characteristics and flaws. When there are no ramifications, the customer will be more apt to change dates, quantities, specifications, ECNs, quality requirements, packaging, etc. These changes create havoc to margins, schedules, and effective use of human resources. When customers have the ability to make changes without consequences, the notification of change may even become delayed. I have witnessed purchasing agents delay notification of changes several weeks, simply because the purchasing agent was busy. With no cost or schedule ramifications, the purchasing agent can operate on his own timetable. I have discussed this issue with my own purchasing agents, and their response is, "it doesn't seem to be a problem for company 'A', so I first notify all organizations who may require an expedite fee or a delivery date change".

When engineers and programmers are required to rework projects due to customer changes, there will be other existing projects with existing commitments that will be negatively affected. Without any additional revenue generated by the account manager from these changes, the margins will be affected. When the account manager fails to manage his customers, these changes multiply and affect the performance of the organization. Strong organizations have procedures for systematically incorporating reasonable customer changes, but every company will experience a domino effect to performance and stress, if this aspect of the business is not controlled. Customer changes create an iceberg effect. There is much more going on underwater than above.

Some organizations, certainly OEMs, will assign a project manager to shepherd large orders through the organization. For some companies the project manager is responsible from concept to installation. This would normally include collecting customer specifications, quoting, engineering, submittals, customer contact, milestone reporting, etc. In other organizations the project manager responsibility is more narrow and begins at the receipt of order and ends at delivery. In these cases the account manager will be less involved in the execution of the order, but still must be involved to some degree in the overall actual management of the order. When there is no project management, the account manager needs to be more involved to facilitate the nebulous details that derail many orders. In fact, frequently the account manager who is the best sales person is not the most successful. Rather, the account manager who is the best project manager is the most successful, because the technical customer appreciates the efficient and organized methodology he provides and that his projects are routinely on time and on budget. This leads to more orders.

OMNIVORES

Most account managers seem to be born either to hunt or to farm. Unfortunately, most account managers do not want to be omnivores—which is what their company requires—but they are determined to be either a herbivore or a carnivore. Many times, the organization needs to shift between hunting and farming. When the schedule is full and there is no available capacity, it is time to be a farmer. When there is available capacity and the business is slow, the account manager better be hunting.

When capacity is full and the carnivore is still hunting, it is frequently the plant manager who must control the hunt. If not, commitments will be made without available capacity, and performance and customers will suffer. One method to slow the hunt is through higher pricing and longer lead times. In essence, if the organization is going to receive orders creating capacity or engineering problems, then accept orders at higher margins that will pay for equipment, overtime, subcontracting, etc.

Hunting while engineering and programming are booked for months into the future and capacity management is displaying workcenters booked for months is not uncommon. The primary driver is multiple account managers pushing for commission. This is not all bad, but it does need to be occasionally controlled. When only some of the account managers are responsible for filling the existing schedule, the remaining account managers desire to meet their corporate or personal targets. Even when they are locked out of the schedule, there will be a desire to short-cycle new orders or heavily persuade that marginal business is strategically important. During this cycle of the business, a leader must be in control who will make wise and tough decisions, or the good times will swiftly move to unhappy customers and lower margins.

Crusty old salesman will say that the customer signs the front of the check. This may be true, but there has to be a balance, so that we do not create a commission rich salesman/happy customer/poor company scenario. The account manager makes more commission, and the customer receives his orders at a low price, while the machining company incurred overtime, expediting, and a lot of stressed-out employees.

Pricing at higher margins when the business level and economy is strong requires a shift away from cost-plus pricing to a market-will-bear pricing approach. When the market is down, manufacturers will be forced to lower prices and lower margins. When the opposite is true, customers may be willing or will be required to pay a higher price to obtain their order within a desired lead time. Remember, customers have customers, so during strong markets there is enough marginal value for shorter lead times or simply to maintain project schedules, so that your higher price gets pushed through the supply chain to the end customer. Do not fret that your higher margin is not fair to your customer. Many distributors obtain a higher margin than machining organizations despite a simplistic business model with much less risk and invested capital than machining organizations. Like farmers or squirrels, the manufacturer needs to put some grain and nuts away for the next long winter–this does not happen by accident, so take the opportunity that the market cycle and your hard work have provided. Winter always comes.

19 SELLING OR ACQUIRING A MACHINING COMPANY

Companies are purchased and sold on a regular basis, and inexpensive capital appears to be maintaining or increasing this trend. It is very beneficial to the machining industry for owners to have the opportunity to sell their business at a fair market value. This will incentivize current owners to invest and grow their business as well as attract new ownership and capital into the industry. The more liquid the assets, the more vitality there will be in the machining industry. When investors have a reliable path for profit, then new talent, intelligence, and energy will be attracted. Whether a business is owned by a second-generation family or short-term investor, it is a positive development that both can be rewarded for their accomplishments.

How does the change of ownership of a business affect the people currently employed at the acquired company? The majority of acquisitions are transacted with the intention to invest and grow the business. This requires a motivated staff, new equipment, and a modern facility. This will create new opportunities and fresh challenges for individuals. When your company or facility is acquired, this is the ultimate stamp of approval. It generally signifies that the team is profitable and creating customer satisfaction. This is business in the 21st century, and all employees need to accept that nobody wants to own a losing facility, which will eventually be shuttered. Conversely, many companies or individuals want to own a profitable facility, and they will pay for the right to invest and enhance the value of the facility and company. Pick, which team you wish to belong to, and move forward.

The recent Warren Buffet acquisition of PCP and his earlier acquisition of toolmaker Iscar have demonstrated legitimate long-term value for pure manufacturing organizations on the largest scale. On the smaller scale, deals are also regularly occurring. The target sector will shift, depending on the macro economy. Medical and energy-oriented manufacturers were desirable in the first decade of this century, and aerospace has been more popular in the second decade. The age of the aircraft fleet, increased fuel economy, and the growth in developing economies have driven new commercial aircraft orders to record numbers in 2014, with deliveries stretched into the next decade due to capacity shortages. Airbus, Boeing, and

Embraer all produced record deliveries in 2015 and entered 2016 with record backlogs. The industry is clamoring for supply chain growth, which has driven selling and acquisitions.

For every buyer there must be a seller. The seller may have received professional consulting to maximize the value of the business or may have spent years massaging the value of the business. The advantage lies with the seller, as they know where the bodies are buried. Achieving the highest value for the business is a combination of timing and good management. It is much wiser to sell when you want to sell rather than when you have to sell. Timing a cyclical economy is more art than science, but after three good years of revenue and profit those considering selling should recognize that their company value is at or near its peak. The market will inevitably turn for the worse, so why delay? It is too easy to be lulled into believing that the economy and/or your business will keep accelerating, but it is just not true.

Besides good timing, high EBIDA, and increasing your multiple, it helps to be lean. Watch the inventory (inventory turns), perform 5S, sign long-term contracts, practice the Machining Pyramid, and improve your metrics as benchmarked against your peers. I am not recommending reducing investment in the business, but it is common for those selling a company to delay avoidable expenses such as facility repairs, software purchases to replace unsupported versions, hiring, and MRO. Likewise, capital purchases have potentially been trimmed or eliminated. Capital equipment acquisitions are required for multiple reasons. The common belief is that new equipment adds capacity to existing processes or provides the capability and technology to create new processes and new offerings to customers. The second component of capital acquisitions—and the more prevailing component—is to replace existing equipment which is fully depreciated and past its useful life (see Figure 19.1). The equipment may be obsolete for either accuracy (Figure 19.2), speed, or—most commonly—lack of available electronic components for maintenance repair. When the normal replacement cycle of either capital or expense items is disrupted, there will be a corresponding spike of equal or greater dollars in the future. In other words, if you are not careful, you will have significant expenditures just to maintain the acquired business because of several years of premeditated financial management by the seller and his advisors.

Acquiring companies will perform due diligence after a letter of intent (LOI) has been signed. The degree of diligence varies between organizations, with the extent of due diligence normally linear to the size of the transaction. Larger organizations tend to have more focus on the synergistic opportunities and profit deltas created by the pending transaction. While smaller companies are focused on revenue growth, larger organizations may be looking for revenue growth as well as to strengthen distribution channels, gain access to new customers, reduce back-office overhead, eliminate a competitor, etc. Much of the due diligence is focused on legal,

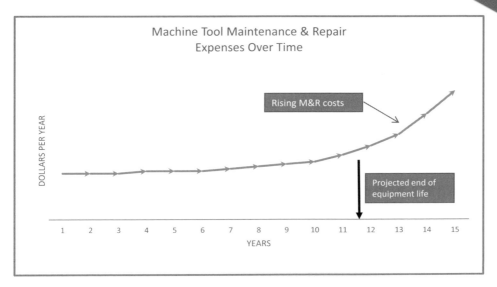

Figure 19.1 Cost-Versus-Age graph – Maintenance and repair costs for a typical machine tool will be flat correcting for inflation over most of ist service life. However, as the service life approaches the equipment's anticipated physical life, maintenance and repairs cost will rapidly increase.

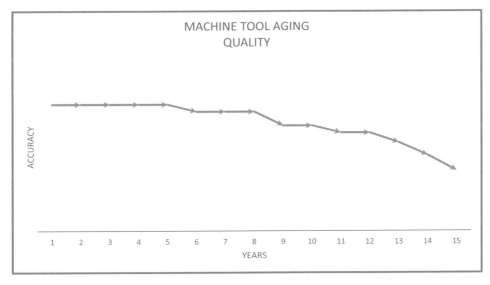

Machine tool aging effects on quality

Figure 19.2 Accuracy-Versus-Age graph

environmental, financial and accounting transactions. Larger organizations are prone to have specific procedures, templates, and metrics for assessment, valuation, and reporting to executives and boards. Are inventories and margins inflated? Do the financial statements accurately reflect the target company's financial condition. Are there complications due to leasing, rental, pension or tax issues, whether past, present, or future? The acquirer will utilize outside legal and financial resources as detectives to determine and quantify risk. Environmental consultants will test and identify abnormalities which will lead to more testing and more consulting. Sales folks will meet with the strategic customers to ascertain any discontent, impending changes, or jeopardy of impact to the business relationship and business levels.

MANUFACTURING ASSESSMENT

While there are ample resources and expenses thrown at redundant analysis of financial, legal, and environmental issues, it is common to undervalue due diligence on the operations side of the business. Executives fall in love with the deal and want to assume that if the margins and accounting appear appropriate, everything else within the target organization must be correct. Operational due diligence includes the manufacturing hardware, facility infrastructure (building and utilities), metrology, quality system, culture, and soft skills. It is not enough to walk through the shop, and it is not enough to contract a used equipment company to provide a cursory value of the equipment. Looking under the hood is a start, but knowing what to look for and the experience to interpret the micro and macro picture is what is required. The manufacturing due diligence can identify mandatory or imminent equipment replacement, facility repairs (compressed air, electrical systems not to code, roofs, underground drainage, etc.) quality gaps, metrology shortcomings, and assessment of the effectiveness of the individual Machining Pyramid building blocks with specific emphasis on processes. Identification of imminent repairs or equipment replacements can provide a six- or seven-figure negotiated reduction in the final price.

The manufacturing due diligence can also identify "soft" aspects of the business. Who are the people within the target organization that create the value? They are not always the highest job titles and they may not be the president's friends and family, whom he would like to see contractually protected and incentivized. What is the value of the due diligence and the acquisition, if you do not retain the personnel who are capable of maintaining value and creating value in the future? Establishing contact with the key personnel to comprehend the structure, flow, and knowledge base of the target is essential. Are there cultural issues that need to be understood and factored?

Pharmaceutical valuations are renowned for determining the strength of the drugs in the development pipeline and not looking in the rear view mirror. Manufacturing and machining organizations may also be currently profitable but not possess the competitive equipment, processes, software, and people to maintain the past performance. Order backlog only tells part of the story. Like pharma, the machining company will have reduced sales when current products are phased out and replaced with products requiring different processes, technologies, accuracies, or price points. What is the development pipeline for the target machining company? An expert can identify these risks. Sometimes the seller knows his business and trends well.

A skilled manufacturing or machining assessment will provide a more thorough and multifaceted grasp of the risk and potential of the target acquisition. If there are multiple targets, the manufacturing assessment may define the optimal choice. The manufacturing assessment will also provide the launching point for integration after the closing.

MANAGING THE ACQUISITION

There is some point after every closing at which the entrepreneurial talent or the key leadership exits the company. This may be day one after the closing, or there may be contracts holding people in place for months or years. Companies are frequently sold either because owners want to exit the business, shareholders want to convert stock to cash, or the performance of the company requires the sale. The end result is either that key people have left immediately or they have received their share of the purchase price and are waiting for their contract to expire. It is rare that the new arrangements provide as much incentive and motivation for the previous working stockholders as existed prior to the sale. Even when the new arrangements are favorable, the key leadership has just received a large payday and will be somewhat distracted. It is not uncommon for the previous ownership to be more concerned with the welfare of their favorite supplier and subordinates than creating value for the new ownership.

New management and new ownership better understand this dynamic very quickly and implement the necessary adjustments to maintain the value of the business. Below is an overview of the three stages of integration.

- Stability: The initial phase after closing is a period lasting from six to twelve months, where the objective is to prove to customers that the new ownership and management structure is committed and capable of meeting all requirements. This will also require stabilizing the workforce, as the highest risk is the loss of key personnel or the loss of a critical mass of personnel, which will jeopardize deliveries, customer confidence, and the ability to train replacements. The primary objective during this time period is to minimize risk until the new management team has firm control.

- Competitors will use the transfer of ownership as an opportunity to steal customers. There is no better time to secure the long-desired customer than during an acquisition. Whether the competitor's sales pitch to the customer is factual, imaginary, or simply unscrupulous, it will occur. Likewise, business competitors and talent competitors will use the same uncertainty to lure key employees. Maintaining a few key employees on contract for a definitive period presents pros and cons. As mentioned earlier, your diligence phase better identify those who really add value. The fundamental problem is that these contracts delay the inevitable. New ownership and new management need to provide an injection of permanence. One method is to complete a capital project, whether that is a new process, new equipment, or a high-profile facility upgrade. This does not have to be excessive in dollars but will help set the tone to customers, visitors, and staff on the commitment of the new team. During times of change, people are highly sensitive to evidence confirming or conflicting the stated direction.

- Transition: After the new management team has achieved stability, the process to realize the strategic goals of the acquisition may begin. This will normally include standardizing financial reporting, HR policies, metrics, organizational structure, and possibly the ERP system. There may be a host of consolidations in support functions such as accounting, purchasing, and sales which are required to provide the ROI for the transaction. During this phase, the management team must articulate a strategy to the workforce that creates a long-term vision of how the new entity will be combined with the parent company to form a stronger entity capable of motivating the A-players to remain engaged and striving for success. In essence, stability has to be maintained throughout the transition phase. This can only be accomplished with buy-in or replacement of key personnel. If the transition is not successful, the organization may revert back to the stability mode.

- Longevity: With the new organizational structure and systems in place, it is time to deliver the ROI and strategic objectives behind the acquisition. This is the time period that demonstrates whether the due diligence and synergistic assumptions were accurate. The project may exceed expectations, and both the acquirer and acquired will be ecstatic. If the results fall short, an autopsy should review whether the plan was bad or the execution was bad (see the Strategy Execution section in the Operations Management chapter).

The ultimate success of the acquisition may require several years to emerge. Often underestimated is the time span of the improvement cycle. Lower-volume companies and OEMs need more time before consolidation and improvements affect the bottom line. High-volume companies yield a short improvement cycle, and the impact to profits occurs quickly. Most consolidations and improvements need three build cycles to hit the bottom line. The first cycle is required to identify or install the improvement. The second cycle is required to debug the improvement, while the third build cycle will harvest the improvement, whether the improvement was cost, quality, or lead time. An OEM or low-volume manufacture may not reproduce the improved product for six months or more. Three cycles would require 18 months. High-volume producers could experience three cycles within a week or within a month.

20 MACHINING TEN COMMANDMENTS

1. Look for similarities, not for differences
2. You are running a sprint, and you are running a marathon
3. Quality is productivity
4. Continual training
5. Make progress, not perfection
6. Inspection must be swift and precise
7. If you are not making chips, you are not making money
8. Always strive for a machinist to operate multiple machines
9. Push decisions to the lowest level
10. Thou shalt not cut air

ELABORATION OF MACHINING TEN COMMANDMENTS

1. **Look for similarities, not for differences:** Engineers, machinists, and other technical people are conditioned through education and training to notice dissimilarities or differences. Learning mass amounts of material and passing tests is accomplished by this method. Profits and robust processes are derived by seeing the glass as half full. When it is time to standardize, implement Lean cells, design new components, develop fixtures, or write CNC programs—it is time for analytical minds to shift gears and find similarities. Why design a new component when one already exists that will fulfill the functional requirements? The volume of a component may not be high enough to justify a Lean cell, but other parts may be processed on the same workcenters creating enough aggregate demand to warrant a cell. When components are 80% similar, take advantage of these traits to utilize existing fixtures, tooling, processes, programs, metrology, etc. and then address the remaining 20%. The difference between an engineer who wants to create something new and fancy versus the engineer who utilizes an existing design is unmeasurable.

2. **You are running a sprint, and you are running a marathon:** Employees seek to achieve customer satisfaction and to optimize metrics. Life becomes a series

of short spurts to ship product today, tomorrow, and to resolve the latest crisis. Loyalty and best intentions can provide near-term results, but who is planning for long-term results? Who is ascertaining future customer needs and developing the people, equipment, software, and processes to provide these needs before the competition? Many organizations are so focused on the short term that they wake up after several years to discover that their equipment and processes are outdated and unproductive.

3. **Quality is productivity:** Scrap and rework directly subtract from the bottom line through the cost of replacement labor, material, and sorting. Borderline quality consumes resources and lowers productivity due to over-inspection and consumption of technical support personnel. Robust processes and quality at the source allow machinists to optimize their time, operate equipment unattended, and operate multiple machine tools.

4. **Continual training:** Train, cross-train, train, cross-train, and train a little more. Quality and productivity are not accidents. Whether the person is a machinist, engineer, or inspector, the transfer of knowledge is simultaneously an investment, insurance policy, and annuity. If you want to be a learning organization with a culture of technological competence and agility, then leadership must set the example and the expectation. Firms that learn faster than their competition will eventually outperform their competition.

5. **Make progress, not perfection:** Greatness is not achieved with a single master stroke. Wars are not won by winning a single battle and with no setbacks. When the expectation is that a new machine or new process will deliver maximum results immediately, or someone is to blame, then the culture becomes risk-adverse and will not be innovative and learning. On the other hand, when progress is recognized and appreciated the team will soar past the original goals. Innovation can frequently be achieved through a series of incremental improvements. Continuous improvement will not be realized without a make-progress-not-perfection mentality.

6. **Inspection must be swift and precise:** Precise is pretty intuitively obvious but still needs elaboration. Besides achieving better quality, the overall machining approach should be to not give away your tolerance through measurement or fixture error. There will be naturally occurring variation generated by machine positioning, tool wear, etc., so it is imperative to eliminate the variation, which should be low-hanging fruit. It is important for inspection to be swift, not only for but so that the machinist or inspector can operate multiple equipment and collect more data. More data via more parts sampled yields better control of inputs, better understanding of the nuances of the component/process, and better centering of the dimension within the tolerance zone, which all leads toward Six Sigma. Swift also refers to timely in that the machinist requires information about his processes immediately to control inputs prior to more parts being produced.

7. **If you are not making chips, you are not making money:** Whether it is grinding, EDM, turning, or another type of machining, the simple goal is for the equipment to be running. The real goal of the entire organization, from the receptionist to the engineer, is to support operations so that the machine tools will be in operation. When you lose time on a machine, you never get it back. If the equipment is not running, customers will never be happy. It is a simple concept, but there are times that many people in the organization lose perspective of the tasks that truly generate the revenue.

8. **Always strive for a machinist to operate multiple machines:** This goal sets an expectation for your engineers, programmers, and machinists to automate, innovate workholding and tooling, and to optimize all metrology requirements. This goal requires robust processes and requires the equipment to run unattended for as long as possible. This goal requires use of available machine accessories for closed-loop machining. All of these mandates are required to successfully compete with developing economies with 10% of the total labor costs.

9. **Push decisions to the lowest level:** This commandment is multi-faceted. You cannot have employees standing around waiting for Oz behind the curtain to tell them what to do or waiting for Oz to answer emails that everyone in the organization is expected to send/copy to Oz. If you expect people to develop and grow, you need to let them make decisions and grow. Will they make a few mistakes? Sure. Their mistakes will be much less significant than the apathy and delay created, if everyone is waiting on a senior manager to make the decision. When people have to seek approval at every step, they cannot act with speed or are likely to just not act.

10. **Thou shalt not cut air:** High-volume manufacturers are more runtime-dependent than low-volume manufacturers, so this commandment is more significant to the high-volume manufacturers. However, everyone needs to adhere to this concept and set the expectation of productivity. Too often the CNC program will have a long approach, travel off the part, or have an excessive retract for clearance. At full speeds with coolant splashing and enclosed cabinets, these issues can be difficult to observe. I have installed GoPro cameras inside machines to identify waste. Additionally, it may be more visible to open the cabinet, turn off the coolant, and videotape a finished part as the program runs at full speed. This permits full visibility to identify waste, brainstorm during a Kaizen event, etc.

Bonus Machining Commandment

11. **Go right before you go fast:** Managers are always in a hurry, whether to meet a customer schedule or to insure productivity. It is beneficial to have a sense of urgency. However, when there is a new process, a new machine, a new program, or just a new setup, things must be right before you can be fast. Ma-

chines and people make bad parts just as fast as good parts. Making them twice is a lot slower and a lot more expensive than spending a little time to be right before you are fast. My organization received a large order for a new family of spinal implants that needed to be delivered within a very short period. This particular implant was an adjustable assembly that required complete process development of eight different sizes. We split the order between two facilities. For the first two weeks I was told how the other plant was already machining parts on multiple shifts and my plant was far behind. I understood the challenges of the project and insured that all machining processes were robust with validated programs, correlated inspection processes, tool change sheets, etc. When we were right, we went fast around the clock and weekends. We delivered great quality and solid margins all ahead of schedule. Meanwhile, the other plant was mired in rework. They went fast before they went right.

THE FIVE COMMANDMENTS THAT WERE LOST

1. Stale data are not actionable.
2. Move people to the work and not the work to the people.
3. Continuous improvement is not an option—it is survival.
4. Always know your cycle times.
5. Continuously strive to use more automation.

21 FUTURE

Western manufacturers are traveling on two parallel paths at the same time. The paths will not cross, so when one bends, it forces the other to divert. The first path is the macroeconomic path forged by trade agreements, developing countries, taxes, exchange rates, and regulations. The second path is the trail we make for ourselves that consists of innovation, connectivity, and the Machining Pyramid. Together, these two paths determine our fate. We have little or no effect on the former, but total control on the latter. If you are in the manufacturing community, you should understand the macroeconomic forces to which you belong. This comprehension provides the necessity and urgency for moving your organization forward on the innovation path. If you, your company, and your social network have a better grasp on the macroeconomic path, perhaps, one day, you will help alter the path.

MACHINING ECOSYSTEM

For the past thirty years there has been a belief that Western economies could flourish, if manufacturing moved to low-cost regions, as innovation would remain in the advanced economies. A 2012 book entitled *Producing Prosperity* by Harvard Professors Gary Pisano and Willey Shih refutes this notion. Pisano and Shih make a convincing argument that when a country loses its ability to manufacture, it loses its ability to innovate. Capacity to solve production problems is as important to the value of innovation as the ability to design the product. Much of today's R&D is either process-embedded innovation or process-driven innovation. According to Piano and Shih, "the movement of manufacturing away from the United States will eventually pull R&D with it" (Pisano and Shih 2012, 70).

Perhaps the most relevant concept to machining discussed in *Producing Prosperity* is the idea of "industrial commons" (ibid., 45). Companies located in certain geographical areas have advantages over others by virtue of their access to the appropriate set of suppliers, workers, engineers, and managerial talent. Obvious examples are semiconductor manufacturers in Taiwan and Singapore, financial services in New York or London, and information technology in the Silicon Valley.

The companies located in these regions form an ecosystem within their industry, with each species contributing resources that benefit others. Machining is integral to an ecosystem that includes engineered raw materials (castings, forgings, exotic metals), heat treaters, coaters, and a customer base including aerospace, medical, energy, and R&D. When machining is strong, the ecosystem is stronger. There is a domino effect in the ecosystem when customers and industries emigrate or evaporate and the entire ecosystem is diluted. Pisano and Shih conclude, "when a (industrial) common erodes it's a deeper and more systematic problem. It means the foundation which future innovation sectors can be built is crumbling" (ibid., 15).

Machining firms need their entire ecosystem to be healthy, if they are going to compete globally. This means that machining companies need other machining companies to be strong to attract and train a talented workforce and a vibrant supplier network. I have presented the Machining Pyramid in the spirit of strengthening the ecosystem.

THE INNOVATION PATH

Machining organizations cannot wait for changes to global competition, reduced trade agreements, or a different political climate. The only insurance is to strengthen your machining company to win regardless of the environment and regardless of the competition. I would begin with assessing your operations in each building block of the Machining Pyramid. Each building block can be separated into its sub-components. If you do not have the impartial and internal expertise necessary to provide a detailed and honest report to form a basis to survive and thrive, get help from the outside. The bar is always moving in machining, so the faster you get started the better.

Currency manipulation, unfair trade practices, and government-owned companies are not what manufacturers want to read and want to write about. Unfortunately, we cannot put our heads in the sand, and we cannot ignore history. Winning is not an accident or coincidence. Winning requires planning and preparation.

The innovation path includes Additive, MTConnnect, machine monitoring, IIoT, Industry 4.0, and all of the yet to be determined technological advancements. This book has emphasized integrating the new connectivity with traditional machining principles and technological improvements. Most companies will chart their own path as they perceive which combination of advancements will provide the greatest impact upon their industry and their business model. I perceive the metrology enhancements (lasers, telecentric lenses, computer inspection, CMMs, and others) integrated with IIOT for real-time feedback to machine controls and intelligent program optimization as the game changer. The exponential increase in measurement volume combined with enhanced data accuracy will generate vast quality and productivity improvements, while freeing critical human resources to add value elsewhere. Quality and reliability will rise with higher CPk, as all features become more centered about the mean. Mega-data will be captured and manipulated, allowing engineers, inspectors, and machinists to innovate and continuously improve. All these activities will drive healthier results throughout the Machining Pyramid.

The ability to gather measurement data immediately after the chips are removed, with high accuracy and in an actionable format, is exceedingly impactful to profits—with or without connectivity to the machine control.

The audience for this book and the machining spectrum in general are very diverse, but there is one common feature. Regardless of your geography or geometry, company size or part size, material type or machine type, high volume or low volume, everyone must have the ability to measure each feature they create and correlate those measurements throughout the internal operations and the external supply chain. Organizations who perform this activity with the most accuracy and productivity are the leaders in their industry.

Success is not an accident, and there is no coasting. The firm that learns and innovates the fastest will be the most agile and will eventually lead. These are the factors that each organization can control.

MACHINING CREATES WEALTH

Machining begins with a $5-chunk of raw material that is transformed by human labor and capital equipment into a $500-component that can be sold to customers the world over. With the $495 of wealth created, we pay employees, suppliers, invest in new equipment, and pay taxes that build schools, roads, and bridges. Manufacturers are high consumers of software, raw materials, accounting & legal services, trucking, and electronics. A medium-sized manufacturing facility will support a number of local and regional suppliers along with a myriad of service and construction personnel. A manufacturing facility is similar to a living, breathing organism. It needs a constant inflow of materials and supplies. It needs to grow, adapt, and change according to changes to its environment. All these activities require expenditures, contractors, consultants, equipment, and hence job creation.

Some industries merely transfer wealth, while other industries create the wealth. The obvious transfer industries are retail and service. The obvious creating industries are mining, farming, and manufacturing. Growth in the manufacturing sector requires more inputs from suppliers, utilities, and mining. Growth in manufacturing requires more services from the finance, legal, I.T. and transportation sectors. A new job created in manufacturing has a higher job multiplier than any other business sector. According to a University of Illinois at Chicago study, "each new manufacturing job is likely to lead to the creation of close to 4.6 additional jobs nationwide" (Scott and Wial 2013, 2). I have seen different studies indicating a range of multipliers, depending on which specific industry created the manufacturing job. Oil and gas or aerospace produce a much higher job multiplier than textile. This is due to the equipment, tooling, purchased components, and supplies that the oil and gas or the aerospace worker will require along with the transportation and services required to move the additional product the worker will create. Regardless of the study or exact multiplier, the consensus is that the manufacturing sector has an oversized impact on the overall economy.

ECONOMIC PRIORITIES

We discussed that we are thirty years into the experiment of sending manufacturing jobs to low-cost countries. How is it going so far? Well, our national debt is 19 trillion and growing rapidly due in part to a higher ratio of low-paying service jobs without healthcare and retirement benefits that replaced manufacturing jobs with healthcare and retirement plans. How is it going so far? Between 1990 and 2011, 70% of the Fortune 500 disappeared from the list. This is a shocking statistic (Brown et al. 2013, 38) and an indication that emigration is not working for corporate America. How many facilities did these 70% close and move and what was the effect on the ecosystem? The job multiplier works in reverse when manufacturing jobs are subtracted. How is it going so far? The U.S. has moved from being the largest creditor nation to being the largest debtor nation.

Since the U.S. decided we did not need to create and build things to generate wealth, what activities and industries did we choose in their place? Some call it financial engineering, but others call it debt, smoke, and mirrors. We first had the savings & loan crisis, then the internet bubble, and finally the housing bubble. These bubbles have removed trillions of dollars of "financially engineered" wealth from the economy.

What are the next bubbles that will remove trillions of dollars of paper wealth? Student loans, Medicare, and unfunded pensions all have financial commitments where there is no realistic solution to the math and all can be future bubbles. What do all the past and future wealth-subtracting bubbles have in common? None of the underlying activities generated enough value to match the commitment. Without a significant level of actual wealth-creating activity in the U.S. (farming, mining, manufacturing, and construction), the economy does not generate enough prosperity to pay for subsidized higher education, pensions, liberal healthcare with no tort restrictions or pharmaceutical limitations, roads, bridges, schools, sewers, and extensive social commitments.

It is quite simple and pretty straightforward—any successful economy must have a meaningful ratio of making, building, and creating. Yes, the world only needs one Switzerland to be the secret banker, and very few nations thrive via tourism. The

U.S. economy is still benefitting from the infrastructure, savings, and pensions produced by the wealth earned by past generations of farmers, miners, manufacturers, and corporations. Without this historical wealth and without the recent low interest rates, our debt and standard of living would be more problematic.

What is going to be the future of Western manufacturing? After decades of ambivalence, everyone from Harvard professors to government think tanks and elected officials are now advocating the need for a strong manufacturing base. The question is where on the hill is the big outsourcing snowball—at the top, middle or bottom? Many in the media would have us believe at the bottom. I expect we are still at the middle of the hill.

GLOBAL COMPETITION

Trade agreements continue to have a negative effect on American manufacturing. Free trade by itself is not negative. The conundrum is that these agreements do not lead to fair trade. American manufacturing is not afraid to compete on a level playing field. The rules of these trade agreements are broken and flouted with no consequence. Issues are rarely arbitrated or taken to the WTO. Even when violations are pursued the damage is irreversible, since years of abuse have occurred—resulting in layoffs and closure of affected companies. Economist and bureaucrats need to realize that these are not just statistics but real people losing their careers, savings, or owners losing their life's work. What is the first hint that a trade agreement may not be in the best interest of the American worker or even American business? When congress includes funding to train displaced (terminated or laid off) workers, it may not be a good trade agreement, and the economic modeling showing that trade gains due to NAFTA are skewed. American corporate shipments of components to their own plants in Mexico for assembly should not be considered an export and should not be considered as a gain under NAFTA. These components are merely assembled and sent back to the U.S., and in reality they are pass-through shipments.

When a foreign government fully supports a chosen industry with a long-term investment and export-tax rebate program in its home country, it is not feasible for private companies in the rest of the world to survive a multi-year onslaught of below market pricing (dumping). Consider the case of Chinese steel selling in the U.S. for half the price as U.S. manufactured steel. Dan Dimicco articulates in his 2015 book, *Made in America*, that steel is not a labor-intensive product, but rather that the cost of steel is driven by raw materials and energy (Dimicco 2015, 218). The U.S. has an advantage over China in both materials and energy. The labor in a ton of steel is less than $10, while the transportation of a ton of steel from China to America costs $40. This means that American steel should have a minimum $30-per-ton advantage over Chinese steel sold in the U.S. The only way that Chinese steel can be sold in the U.S. is massive support from the Chinese government.

What will happen to international steel prices when Western steel manufacturing succumbs to this long-term onslaught?

The primary reason why I believe that the snowball is not at the bottom of the hill is the still massive difference in labor cost. A 2015 Congressional Research Service report (Levinson, *U.S. Manufacturing in International Perspective*) lists total hourly compensation costs (including benefits) for the U.S. at $36.34–at $6.62 for Mexico, at $3.07 for China, and at $1.57 for India (see Figure 21.1). The media likes to report rising wages in China. While this may be true, you must consider that a 10% increase in China is $.31, while a 2% increase in the U.S. is $.73. The current ratio of U.S. labor compensation to Chinese labor compensation is 11.8 to 1. Even with a ten-year projection of 2% growth in the U.S. and 10% growth in China the total compensation gap will increase from the current $33.27 to $36.19 per hour.

WAGE RATIOS							
YEAR	INDIA	U.S/INDIA RATIO	CHINA	U.S./CHINA RATIO	MEXICO	U.S./MEXICO RATIO	U.S.
*1	$1.59	22.86	$3.07	11.84	$6.62	5.49	$36.34
2	$1.75	21.19	$3.38	10.98	$7.28	5.09	$37.07
3	$1.92	19.65	$3.71	10.18	$8.01	4.72	$37.81
4	$2.12	18.22	$4.09	9.44	$8.81	4.38	$38.56
5	$2.33	16.90	$4.49	8.75	$9.69	4.06	$39.34
6	$2.56	15.67	$4.94	8.11	$10.66	3.76	$40.12
7	$2.82	14.53	$5.44	7.52	$11.73	3.49	$40.92
8	$3.10	13.47	$5.98	6.98	$12.90	3.24	$41.74
9	$3.41	12.49	$6.58	6.47	$14.19	3.00	$42.58
10	$3.75	11.58	$7.24	6.00	$15.61	2.78	$43.43

* Source: 2015 Congressional Research Service
Assumptions for years 2-10:
  2% ANNUAL U.S. GROWTH, 10% GROWTH FOR INDIA, CHINA, AND MEXICO

Figure 21.1

Is the Chinese market open to American exports? The largest exports by dollar value from America to China are farm products and scrap metals. Neither of these really create any jobs. Of the top ten U.S. trading partners, China receives the lowest exports per capita by a substantial margin. Japan is the next lowest at six times greater than China, Belgium 36 times greater, and Canada 93 times greater (data source: U.S. Census Bureau, 2015). Meanwhile, China's exports to the U.S. increased from 2014 to 2015 for the 29th time in the last thirty years (2009 is the exception) by $15 billion. The $15 billion is not just a number but should be viewed as 1500 U.S. companies producing $10M less in revenue. Because the U.S. economy has experienced very low growth since 2008, the zero-sum perspective is applicable. The amount of import trade dollars into China is not an accident. A massive industrial complex such as China should be ripe for industrial equipment and technology from the U.S.

These statistics are only part of the story. Chinese officials may further manipulate currency to compensate for any wage increases to protect cost advantages. Data and discussions on Chinese wage growth focus on the eastern coastal area. Central and western China still have a much lower wage structure and much less upward pressure. China continues to develop infrastructure away from coastal areas, and manufacturers will continue to move inland.

Does wage growth in China matter, if currency manipulation continues on a massive scale? Consider the history of the yen during Japan's industrial ascendancy from 1970 to 1995. The yen moved from 350 versus the dollar in 1970 to approximately 80 versus the dollar in 1995. This is more than 400% strengthening (see Figure 21.2). However, during a similar 25-year period of the Chinese industrial ascendancy from 1990 to 2015 the Chinese renminbi actually weakened against the dollar from 4 renminbi per dollar in 1990 to just over 6 renminbi per dollar in 2015 (see Figure 21.3). How does a currency weaken more than 50% during a period of accumulating more than $3.5 trillion in foreign currency reserves and achieving more than $350 billion in annual foreign trade surplus with the U.S.? The $350 billion U.S. trade deficit with China is the largest in the history of the world. Dimicco emphasizes that we have been in a trade conflict with China since the mid-1990s and that Chinese currency manipulation has cost the U.S. more than 2.1 million jobs (Dimicco 2015, 100). The problem is that the U.S. is not responding to these aggressive and mercantilist trade practices. The behavior is similar to the Aztecs facing Cortez, rather than a nation understanding that its wealth, security, and future is at stake.

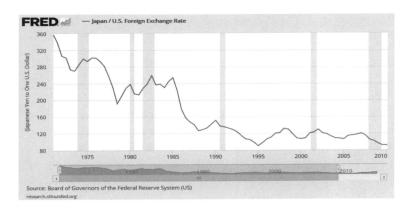

Source: Board of Governors of the Federal Reserve System (US)
research.stlouisfed.org

Figure 21.2 Japanese Exchange Rate

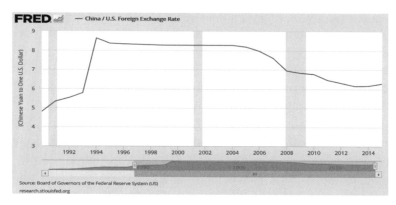

Figure 21.3 Chinese Exchange Rate

Compounding unfair trade practices is the natural progression of a developing nation's growth within skill and knowledge industries. Their machining capabilities are gradually increasing, partly because of the rapid movement and technology sharing of Western companies. Continuous improvement in developing nations includes better machine tools, processes, and skills. Figure 21.4 displays the trend of the developing countries' capabilities related to complexity and volume.

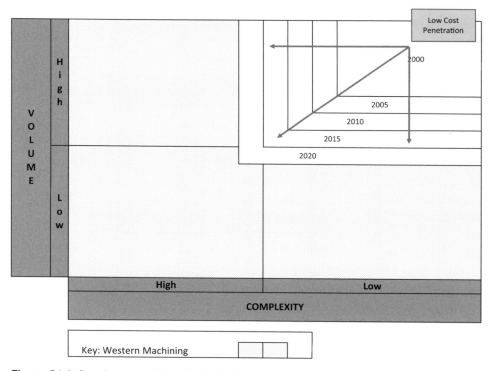

Figure 21.4 Development of Low-Cost Machining

Western companies should invest in developing countries and source products in developing countries. It is acceptable for organizations to seek to increase shareholder value by expanding markets and sourcing lower-cost materials. There will be some natural migration of jobs to lower-cost regions, based on the competitive advantage principle articulated in Adam Smith's centuries old classic *Wealth of Nations*. However, the international trade community should insist on rules-based free trade, enforce the rule of law, and not permit one country to unfairly benefit while creating an economic crisis in another country.

Creative destruction has merit when a new industry is created and an old industry is destroyed, for example, when the creation of the automotive industry destroyed the stage coach industry, when the cellular industry destroyed the home telephone industry, or when the telephone originally destroyed the telegraph. In these cases, the new jobs replaced the old. America has forgotten the creative portion of creative destruction. We destroy entire industries by sending them abroad at a speed that the level of investment, creativity, and old-fashioned need is not able to replace.

Let us summarize the challenges facing American industry:

- Currency manipulation against WTO rules, making foreign products cheaper
- Trade agreements that are not enforced
- Foreign governments subsidizing their corporations for exports
- Foreign government ownership of corporations
- High effective U.S. corporate tax rates
- Regulatory environments creating competitive handicaps
- Total compensation ratios for employees greater that 10:1

These are a lot of boulders to put in the backpack. The weight is enough to bring even the best companies to their knees.

WESTERN CONSUMER MORAL RESPONSIBILITY

We all deserve and want a safe work environment, clean rivers, and clean water. The problem is not Western regulatory agencies but the lack of self-regulation or government regulation in the developing countries. The regulatory and behavioral gap exacerbates the compensation gap between developed and developing nations. Western consumers would not purchase a product manufactured in their own country from a company with unsafe working conditions, underage labor, and a history of polluting. Yet we shop daily at retailers that import products from overseas regions that routinely employ these practices. The most egregious practices occur at the tier two, three, and four level, which are hidden from the retailer and end customer. Are we not hypocrites when we prevent unsafe conditions in our own backyard but openly use our wallets to reinforce and expand these practices in other regions of the globe?

One solution is to demand that our retailers tag every product with a verifiable Manufacturer's Identification Number (MIN) that would allow potential consumers to view online a minimum set of criteria for each MIN as shown below:

- Internal and external photos of originating facility
- Compensation and benefit data
- Current and past company name
- Tier 2 and tier 3 suppliers and their MIN
- Worker accident records
- Address
- Longitude and latitude coordinates

We desire jobs to be available globally, but do we not also have a moral obligation to encourage our supplier (YES, when we shop we are the customer, and the retailer is our supplier) to partner with only those manufacturers who provide a minimal level of safe and pollution-free work environment? Coming from a society where "sustainability, "recycled", and "free range" are demanded by the consumer, should the improvement of other people's health and the elimination of poverty not also be an aspiration?

Henry Ford substantially created the U.S. middle class by increasing the pay of his workers from $2.38 per day to $5 per day. This single event launched the purchasing power of the U.S. middle class that created a consumer market responsible for increasing the standard of living for millions of Americans as well as foreigners for generations.

Can we not provide an incentive for world-wide manufacturers to sell more products through enhancing their work environment and employee compensation? Can we not help to foster a meaningful middle class in developing countries? A middle class capable of purchasing and consuming products to enable economic growth in a manner other than exporting?

TARGETED R&D

The U.S. has shed more than six million manufacturing jobs since 2000 primarily to low-cost regions. Germany, a country with higher wages and very high social cost has not lost any manufacturing jobs during this period. While there is no standard definition for a high-technology industry, it is generally understood that German manufacturing contains a high ratio of high-tech value-added manufacturers. The German Fraunhofer Institute fosters R&D through joint private and government funding at 66 different locations throughout the world, including the U.S. The Fraunhofer Institute employs more than 24,000 people with an annual research budget of 2 billion Euro (Fraunhofer.org). The United States has no such program in scope or funding. Singapore, a country equivalent in size and population to Chicago, outspends the U.S. in R&D for industrial production and technology.

While the U.S. possesses the largest total R&D investment of any nation in total dollars, they are tenth as a percent of GDP. Very little of these funds are invested in commercially oriented industrial R&D fields. The "German and South Korean governments invest nearly nine times more than the U.S. in R&D for industrial production and technology, while Japan spends four times more" (Atkinson 2014). Atkinson stresses that all three of these nations continue to have a large manufacturing trade surplus.

Many individuals prefer less government involvement. It must be recognized that U.S. manufacturers face a consortium of foreign competitors financed by domination in their home market and financed directly or indirectly by their governments. Industries that have been nearly destroyed are machine tools, semiconductors, and steel.

Yes, to stem the tide will take cooperation between private industry, academia, and government. The Apollo missions were an example of a private/public partnership that yielded long-term success. If R&D spending levels are not increased, then at a minimum a tax policy more favorable to manufacturers and manufacturing R&D will be needed.

There will always be questions how we spend limited government resources. When do we need to provide fish, and when do we need to teach people how to fish? Well, you need a lake to do either of these. High-tech manufacturing is one of the few sectors that can create the volume of jobs, the wealth, and the job multiplier to have an impact on an economy the size of the U.S. economy. Investment in industrial production, industrial technology, and commercialization of new products create and expand the lakes.

The lake is the place for the fisherman to go to work, and the lake is the place to extract taxes that pay for the fish. The more people fish, the less fish we will have to provide. The key is to insure that we maintain a thriving lake. This requires us all to work together on the ecosystem of the lake!

BIBLIOGRAPHY

Ackermann, R. 2013. "M2M, Internet of Things & Industry 4.0: An Industry Perspective." Second Indi-an-German Workshop, January 2013. *http://www.acatech.de/fileadmin/user_upload/Baumstruktur_ nach_Website/Acatech/root/de/Material_fuer_Sonderseiten/Second_German-Indian-Workshop/India_ 01_13_Industrie40_m2m_Ackermann_SAP.pdf*

Albert, M. 2015a. "Bearing Down on Industry 4.0." *Modern Machine Shop* (Dec.): 74–81.

Albert, M. 2015b. "Seven Things to Know about the Internet of Things and Industry 4.0." *Modern Machine Shop* (Sept.): 82–87.

Ames, M. et al. 2011. "Quality Management Systems: The 10 Most Common Myths." *Quality Management Digest*, 10 August 2011. *http://www.qualitydigest.com/inside/quality-insider-column/quality-manage ment-systems-10-most-common-myths.html#*

Anderson, J. 2015. "Manufacturing Leads Internet of Things with MTConnect Communications Stand-ard." Engineering.com, 10 February 2015. */www.engineering.com/AdvancedManufacturing/ ArticleID/9573/Manufacturing-Leads-Internet-of-Things-with-MTConnect-Communications-Standard. aspx*

Atkinson R. 2014. "Government R&D and Manufacturing Competitiveness." *The Hill*, 07 July 2014.

Ahlstôm, J. 2015. "How to Succeed with Continuous Improvement: A Primer for Becoming the Best in the World." *Cutting Tool Engineering* 67, no. 8 (Aug.): 13–14.

Baden, B. 2013. "What America Exports to China." *China Business Review*, 15 May 2013. *http://www. chinabusinessreview.com/is-china-a-market-economy-it-doesnt-matter/*

Bhote, K. 1991. *World Class Quality: Using Design of Experiments to Make It Happen.* New York: Amacom.

Bond, S. 2012. "Lean Manufacturing Through Compact Machining Cell Advancements." *Manufacturing Engineering. http://www.sme.org/MEMagazine/Article.aspx?id=68210&taxid=1415*

Bossidy, L. and R. Charan. 2002. *Execution: The Discipline of Getting Things Done.* New York: Crown Business.

Bowman, R. 2014. "This Year's Recall 'Pileup' is a Supply-Chain Nightmare for Automakers." *Forbes*, 10 June 2014.

Bradberry, T. 2015. "Are You a Follower or a Leader." *Forbes*, 18 August 2015. *http://www.forbes.com/ sites/travisbradberry/2015/08/18/are-you-a-leader-or-a-follower/#5acccd146a55*

Brooks, R. 2012. "New Adaptive Control CNC." *American Machinist. http://americanmachinist.com/ automation/new-adaptive-control-cnc*

Brothers, E. 2016. "2016 Aerospace Forecast." *Aerospace Manufacturing and Design. http://magazine. onlineamd.com/article/february-2016/2016-aerospace-forecast.aspx?platform=hootsuite*

Brown, S. et al. 2013. *Strategic Operations Management* (3rd ed.). London: Routledge.

Brue, G., and R. Howes. 2006. *Six Sigma.* New York: McGraw-Hill.

CAPA. 2014. "Record Airline Fleet Order Backlog has Winners and Losers." *CAPA Centre for Aviation.* *http://centreforaviation.com/analysis/record-airline-fleet-order-backlog-has-winners–losers—and-questions-about-how-it-will-be-funded-152713*

Case, B. 2013. "Automakers Spur $3 Billion Boom for Made-in-Mexico Steel." *Bloomberg,* 17 April 2013.

Chaneski, W. 2014. "Commit to Success." *Modern Machine Shop* 87, no. 6 (Nov.): 40–42.

Chaneski, W. 2015. "Benefits of a Kaizen Event." *Modern Machine Shop* 87, no. 10 (March): 34–36.

Christman, A. 2002. "Digital Manufacturing: An Emerging Technology" *Modern Machine Shop. http://www.mmsonline.com/columns/digital-manufacturing-an-emerging-technology*

Cohen, S., and J. Roussel. 2013. *Strategic Supply Chain Management: The Five Disciplines for Top Performance* (2nd ed.). New York: McGraw-Hill.

Conley, G. 2015. "Industry 4.0 Creates the Vision for the Smart Factory." *Cincinnati Business Journal,* 30 March 2015.

Conner, C. 2009. *Lean Manufacturing for the Small Shop* (2nd ed.). Dearborn, Mi.: Society of Manufacturing Engineers.

Danjou, C. et al. 2015. "Closed-loop Manufacturing, A STEP-NC Process for Data Feedback: A Case Study." In *Proceedings of the 48th CIRP Conference on Manufacturing Systems,* Vol. 41: 852–857. *http://www.sciencedirect.com/science/article/pii/S2212827115011130*

Davis, J. 2015. "Smart Manufacturing for Optimum Industry Health." *Manufacturing Engineering* 155, no. 3 (Sept.): 16.

Dimicco, D. 2015. *American Made: Why Making Things Will Return US to Greatness.* New York: Palgrave Macmillan Trade.

Elliott, M. 2013. "Global Steel 2013, A New World, A New Strategy." Ernst & Young *http://www.ey.com/IN/en/Industries/Mining—Metals/Global-steel-2013—Overview*

Frisbie, J. and E. Ennis. 2016. "Is China a "Market Economy"? It Doesn't Matter." *China Business Review,* 26 April 2016. *http://www.chinabusinessreview.com/is-china-a-market-economy-it-doesnt-matter/*

Giesecke, J. and B. McNeil, B. 1999. *Core Competencies and the Learning Organization.* Lincoln: University of Nebraska. *http://digitalcommons.unl.edu/libraryscience/60*

Gilles, C. 2006. "New Competitive Realities in Steel." *Millennium Steel. http://www.steelconsult.com/ArticleSteelConsultinMillenniumSteel_English.pdf*

Goldratt, E. 1984. *The Goal.* Great Barrington, MA.: North River Press.

Hall, D. 2008. "Five Ways to Increase Profit Margins." *Bloomberg Businessweek,* 22 August 2008.

Hanson, K. 2015. "Leaning Towards Lean: A Journey of Waste Reduction and Continuous Improvement for Machine Shops." *Cutting Tool Engineering* 67, no. 2 (Feb.): 68–75.

Hartman, B., King, W. and S. Narayanan. 2015. "Digital Manufacturing: The Revolution will be Virtualized." McKinsey&Company. *http://www.mckinsey.com/insights/operations/digital_manufacturing_the_revolution_will_be_virtualized*

Huang, Q., Zhou, S. and S. Jianjun, S. 2002. "Diagnosis of multi-Operational Machining Through Variation Propagation Analysis." In *Robotics and Computer Integrated Manufacturing, 11th International Conference on Flexible Automation and Intelligent Manufacturing,* Vol. 18, nos. 3–4: 233–239.

Huechemer, B. 2016. "Let's Set the Record Straight on Industry 4.0." *Modern Machine Shop* 88, no. 8 (Jan.): 42–44.

Idhammar, T. 2016. "Cost and Estimated Replacement Value." *ReliablePlant. http://www.reliableplant.com/Read/7050/cost-replacement-value*

Iverson, J. 2016. "Bridging the Skills Gap: Where's the Good News?" 20 January 2016. *http://www.mmsonline.com/cdn/cms/Skills_shortage.pdf*

Jennings, K. 2015. "Offshoring Again." *Cutting Tool Engineering* 67, no. 10 (Oct.): 24.

Johnson, S. 2007. "The Role of Sales Engineer in Technical Sales." *Pragmatic Marketing. http://pragmat icmarketing.com/resources/the-role-of-sales-engineer-in-technical-sales*

Kann, J. 2015. "SME Speaks: Transforming Manufacturing through the Connected Enterprise." *Manufacturing Engineering* 155, no. 3 (Sept.): 21–22.

Kamphake, J. and B. Wilson. 2015. "Pardon the Disruption!" *The Cincinnati Business Journal*, 30 March 2015. *http://www.bizjournals.com/cincinnati/feature/technology/pardon-the-disruption.html*

Koenig, B. 2015. "Lean's Next Act is a Balancing Act." *Manufacturing Engineering* 155, no. 2 (Aug.): 78–83.

Kokenmuller, N. 2015. "Technical & Functional Skill for Sales People." *Demand Media. http://work. chron.com/technical-functional-skills-sales-people-22190.html*

Korn, D. 2015. "Best Practices of Top U.S. Shops." *Modern Machine Shop* 88, no. 3 (Aug.): 70–77.

Korn, D. 2015. "All-in for Apprentices." *Modern Machine Shop* 87, no. 12 (May): 20–22.

Kotter, J. 1996. *Leading Change*. Boston: Harvard Business Review Press.

Kownatzki, C. 2010. "Strong-arming the Yuan is More Difficult Than You Realize." *Business Insider. http://www.businessinsider.com/yuan-dollar-conflict-2010-4*

Krotkin, J. and M. Shires. 2015. "The Cities Leading A U.S. Manufacturing Revival." *Forbes*, 23 July 2015. *http://www.forbes.com/sites/joelkotkin/2015/07/23/the-cities-leading-a-u-s-manufacturing-revival/ #5570a4951d0a*

Kruse, K. 2015. "Focus on the Vital Few for Disciplined Leadership." *Forbes*, 24 October 2015. *http:// www.forbes.com/sites/kevinkruse/2015/10/24/the-disciplined-leader-john-manning/#1e4376782498*

Lafley, A. and R. Martin. 2013. *Playing To Win: How Strategy Really Works*. Boston: Harvard Business Review Press.

Linke, B. et al. 2016. Workshop: U.S.-Germany Collaborative Research in Advanced Manufacturing in Darmstadt, Germany, February 2016. *www.sme.org/uploadedFiles/NAMRI/German_US_Interactions_ Final.docx*

Levinson, M. 2015. "U.S. Manufacturing in International Perspective." *Congressional Research Service*, 17 March 2015.

Magalhães, L. and I. Kanban. 2015. LinkedIn, 15 May 2015. *https://www.linkedin.com/pulse/kanban-ivan-luizio-magalh%C3%A3es*

Mayberry, D. 2005. "Economic Wealth: A Three-Step Process." *Federal Reserve of Minneapolis*, September 2005. *https://www.minneapolisfed.org/publications/the-region/economic-wealth-a-threestep-process*

Mayeda, A. 2015. "IMF Approves Reserve Currency Status for China's Yuan." *Bloomberg News*, 30 November 2015.

Miller, J., and L. Beilfuss, L. 2015. "U.S. Steelmakers Seek Antidumping Action Against China, Four Others." *Wall Street Journal*, 3 June 2015.

Morey, B. 2009. "Feeding the Loop." *Manufacturing Engineering. http://www.sme.org/MEMagazine/ Article.aspx?id=27225&taxid=1424*

Meyer, C. 1993. *Fast Cycle Time: How to Align Purpose, Strategy, and Structure for Speed*. New York: The Free Press.

Nemeth, D. 2015. "Bridging Gaps: Manufacturing consultant leverages CAM software to help customers adopt transformative technology." *Cutting Tool Engineering. http://www.ctemag.com/news-videos/ articles/bridging-gaps*

Nithyanandam, G, and R. Pezhinkattil. 2014. "A Six Sigma Approach for Precision Machining in Milling." *Science Direct* 97: 1474–1488. *http://www.sciencedirect.com/science/article/pii/S187770581403 4997*

Ogewell, V. 2014. "Siemens, Tesis PLMware and Industry 4.0." Engineering.com, 14 January 2014. *http://www.engineering.com/PLMERP/ArticleID/6958/Siemens-Tesis-PLMware-and-Industry-40.aspx*

Pande, P. et al. 2000. *The Six Sigma Way: How GE, Motorola, and Other Top Companies Are Honing Their Performance.* New York, McGraw-Hill.

Pande, P. et al. 2014. *The Six Sigma Way: How to Maximize the Impact of Your Change and Improvement Efforts* (2nd ed.). New York: McGraw-Hill.

Pisano, G. and C. Shih, 2012. *Producing Prosperity: Why America Needs a Renaissance.* Boston: Harvard Business School.

Porter, M. 1980. *Competitive Strategy: Techniques For Analyzing Industries And Competitors.* New York: The Free Press.

Saez, E. 2004. "Income and Wealth Concentration in a Historical and International Perspective." UC Berkley and NBER, 21 February 2004. *http://eml.berkeley.edu/~saez/berkeleysympo2.pdf*

Scheyder, E. 2015. "Hess is mirroring Toyota Manufacturing Process to Slash Well Costs." reuters.com, 4 February 2015. *http://www/reuters.com/article/2015/04/02/dugcongerence-hess*

Schonberger, R. 1986. *World Class Manufacturing: The Lessons of Simplicity Applied.* New York: The Free Press.

Schonberger, R. 2008. *Best Practices in Lean Six Sigma Process Improvement: A Deeper Look.* Hoboken, NJ: John Wiley & Sons Inc.

Scott, E. and H. Wial. 2013. *Multiplying Jobs: How Manufacturing Contributes to Employment Growth in Chicago and the Nation.* Chicago: University of Illinois at Chicago Center for Urban Economic Development.

Seidle, E. 2013. "The Greatest Retirement Crisis in American History." *Forbes,* 20 March 2013. *http://www.forbes.com/sites/edwardsiedle/2013/03/20/the-greatest-retirement-crisis-in-american-history/#54092ed71b88*

Shoshanah, C. and Joseph Roussel. 2013. *Strategic Supply Chain Management: The Five Disciplines for Top Performance* (2nd ed.). New York: McGraw-Hill.

Smil, V. 2013. *Made In The USA: The Rise And Retreat Of American Manufacturing.* Cambridge, MA: MIT Press.

Standard, C. and D. Davis. 1999. *Running Today's Factory: A Proven Strategy for Lean Manufacturing.* Cincinnati: Hanser Publications.

Tate, C. 2015. "Lean Drives Modularity." *Cutting Tool Engineering* 67, no. 10 (Oct.): 34–36.

Tisdall, N. 2015. "Industry Trends." Scotiabank, April 2015. *www.scotiabank.com*

Trapp, R. 2015. "Why Charisma Beats Performance in the Leadership Race." *Forbes,* 30 November 2015. *www.forbes.com*

Vijayaraghavan, A. 2015. "The Human-in-the-Loop." *Modern Machine Shop* 88, no. 3 (Aug.): 34–46.

Webster, S. 2015. "Inside America's Bold Plan to Revive Manufacturing." *Manufacturing Engineering* 154, no. 6 (June): 49.1–49.15.

Westcott, R. 2005. "Corrective vs. Preventive Action." *Quality Progress. http://asq.org/quality-progress/2005/03/problem-solving/corrective-vs-preventive-action.html*

Womack, J. et al. 2007. *The Machine that Changed the World: The Story of Lean Production.* New York: Free Press.

Zelinski, P. 2008. "Machining Megatrends." *Modern Machine Shop. http://www.mmsonline.com/columns/machining-megatrends*

Zelinski, P. 2010. "Modern Manufacturing in 12 Tweets." *Modern Machine Shop. http://www.mmsonline.com/columns/modern-manufacturing-in-12-tweets*

Zelinski, P. 2015. "Bring Your Questions." *Modern Machine Shop* 88, no. 3 (Aug.): 20.

Index